Angelika Bikner-Ahsbahs
Andreas Vohns · Regina Bruder
Oliver Schmitt · Willi Dörfler

Theories in and of Mathematics Education

Theory Strands in German Speaking Countries

Angelika Bikner-Ahsbahs
Faculty of Mathematics and Information Technology
University of Bremen
Bremen
Germany

Andreas Vohns
Department of Mathematics Education
Alpen-Adria-Universität Klagenfurt
Klagenfurt
Austria

Regina Bruder
Fachbereich Mathematik
Technical University Darmstadt
Darmstadt, Hessen
Germany

Oliver Schmitt
Fachbereich Mathematik
Technical University Darmstadt
Darmstadt, Hessen
Germany

Willi Dörfler
Institut für Didaktik der Mathematik
Alpen-Adria-Universität Klagenfurt
Klagenfurt
Austria

DOI 10.1007/978-3-319-42589-4

Library of Congress Control Number: 2016945849

Contents

Chapter 1
Introduction

Angelika Bikner-Ahsbahs

In the 1970s and 1980s, mathematics education was established as a scientific discipline in German-speaking countries through a process of institutionalization at universities, the foundation of scientific media, and a scientific society. This raised the question of how far the didactics of mathematics had been developed as a scientific discipline. This question was discussed intensely in the 1980s, with both appreciative and critical reference to Kuhn and Masterman. In 1984, Hans-Georg Steiner inaugurated a series of international conferences on Theories of Mathematics Education (TME), pursuing a scientific program aimed at founding and developing the didactics of mathematics as a scientific discipline. Chapter 2 will show how this discussion was related to a discourse on theories. Chapters 3 and 4 will present two theory strands from German-speaking countries: with reference to Peirce and Wittgenstein, semiotic approaches are presented by Willi Dörfler and a contribution to activity theory in the work of Joachim Lompscher is presented by Regina Bruder and Oliver Schmitt.

Addressing some TME issues, a more bottom-up meta-theoretical approach is investigated in the networking of theories approach today. Chapter 5 will expound this approach and its relation to the TME program. In this chapter, the reader is also invited to take up this line of thought and pursue the networking of the two presented theoretical views (from Chaps. 3 and 4) in the analysis of an empirical case study of learning fractions and in an examination of how meta-theoretical reflections may result in comprehending the relation of the two theories and the complexity of teaching and learning better. In Chap. 6, we will look back in a short summary and look ahead, proposing some general issues for a future discourse in the field.

A. Bikner-Ahsbahs (✉)
Faculty of Mathematics and Information Technology, University of Bremen,
Bibliothekstrasse 1, 28359 Bremen, Germany
e-mail: bikner@math.uni-bremen.de

A. Bikner-Ahsbahs et al., *Theories in and of Mathematics Education*,
ICME-13 Topical Surveys, DOI 10.1007/978-3-319-42589-4_1

Finally, a list of references and a specific list for further reading are offered. Since this survey focuses mainly on the German community of mathematics education, the references encompass many German publications.

Chapter 2
Theories in Mathematics Education as a Scientific Discipline

Angelika Bikner-Ahsbahs and Andreas Vohns

This first chapter of the survey addresses the historical situation of the community of mathematics education in German-speaking countries from the 1970s to the beginning 21st century and its discussion about the concept of theories related to mathematics education as a scientific discipline both in German-speaking countries and internationally.

2.1 How to Understand Theories and How They Relate to Mathematics Education as a Scientific Discipline: A Discussion in the 1980s

On an institutional and organizational level, the 1970s and early 1980s were a time of great change for mathematics education in the former West Germany[1]—both in school and as a research domain. The Institute for Didactics of Mathematics (*Institut für Didaktik der Mathematik*, IDM) was founded in 1973 in Bielefeld as the first research institute in a German-speaking country specifically dedicated to mathematics education research. In 1975 the Society of Didactics of Mathematics

[1]For an overview including the development in Austria, see Dörfler (2013b); for an account on the development in Eastern Germany, see Walsch (2003).

A. Bikner-Ahsbahs (✉)
Faculty of Mathematics and Information Technology, University of Bremen, Bibliothekstrasse 1, 28359 Bremen, Germany
e-mail: bikner@math.uni-bremen.de

A. Vohns
Department of Mathematics Education, Alpen-Adria-Universität Klagenfurt, Sterneckstraße 15, 9020 Klagenfurt, Austria
e-mail: andreas.vohns@aau.at

A. Bikner-Ahsbahs et al., *Theories in and of Mathematics Education*, ICME-13 Topical Surveys, DOI 10.1007/978-3-319-42589-4_2

(*Gesellschaft für Didaktik der Mathematik*, GDM) was founded as the scientific society of mathematics educators in German-speaking countries (see Bauersfeld et al. 1984, pp. 169–197; Toepell 2004).

The teachers' colleges (*Pädagogische Hochschulen*), at that time the home of many mathematics educators, were either integrated into full universities or developed into universities of education that were entitled to award doctorates. The Hamburg Treaty (*Hamburger Abkommen*, KMK 1964/71) adopted in 1964 by the Standing Conference of Ministers of Education and Cultural Affairs (KMK) led to considerable organizational changes within the German school system. The traditional *Volksschule* (a common school covering both primary and secondary education, Grades 1–8) was abolished and led to an even more differentiated secondary school system, establishing two types of secondary schools called *Hauptschule* and *Realschule* in addition to the already established *Gymnasium*. The Hamburg Treaty also abolished the designations of the school subjects dedicated to mathematics education, which was traditionally called *Rechnen* (translates as "practical arithmetic") in the *Volksschule* and as *Mathematik* in *Gymnasium* (see Griesel 2001; Müller and Wittmann 1984, pp. 146–170).

Likewise, there was a strong interest in discussing how far mathematics education had developed as a scientific discipline, as documented in both of the German-language journals on mathematics education founded at that time: the *Zentralblatt für Didaktik der Mathematik* (*ZDM*, founded in 1969) and the *Journal für Mathematik-Didaktik* (*JMD*, founded in 1980). In these discussions, two main aspects were addressed: the role and suitable concept of theories for mathematics education and how mathematics education as a scientific discipline was to be founded and could be further developed. However, both aspects are deeply intertwined.

Issue 6 (1974) of *ZDM* was dedicated to a broad discussion of the current state of the field of "Didactics of Mathematics"/mathematics education. The issue was edited by Hans-Georg Steiner and included contributions from Bigalke (1974), Freudenthal (1974), Griesel (1974), Otte (1974), and Wittmann (1974), among others. These articles were focused around the questions of (1) how to conceptualize the subject area or domain of discourse of mathematics education as a scientific discipline, (2) how mathematics education may substantiate its scientific character, and (3) how to frame its relation to reference disciplines, especially mathematics, psychology, and educational science. While there has been a great diversity in the approaches to these questions and, likewise, to the definitions of "Didactics of Mathematics" given by the various authors, cautioning against reductionist approaches seemed to be a common topic of these papers. That is, the authors agreed upon the view that mathematics education cannot be meaningfully conceptualized as a subdomain of mathematics, psychology, or educational science alone.

The role of theory was more explicitly discussed about 10 years later in two papers (Burscheid 1983; Bigalke 1984) and in two comments (Fischer 1983; Steiner 1983) published in the *JMD*. As an example of the discussion about theory at that time, we will convey the different positions in these papers in more detail.

In 1983, Burscheid used the model of Kuhn and Masterman (see Kuhn 1970; Masterman 1970) to explore the developmental stage of mathematics education as a scientific discipline. He justified this approach by claiming that every science represents its results through theories and therefore mathematics education as a science is obliged to develop theories and make its results testable (Burscheid 1983, p. 222). The model of Kuhn and Masterman describes scientific communities and their development using paradigms. By investigating mainly natural sciences, Kuhn has characterized a paradigm by four components:

1. *Symbolic generalizations*: "expressions, deployed without question or dissent…, which can readily be cast in a logical form" (Kuhn 1970, p. 182) or a mathematical model: in other words, scientific laws, e.g., Newton's law of motion.
2. *Metaphysical presumptions*: as faith in specific models of thought or "shared commitment to beliefs," such as "heat is the kinetic energy of the constituent parts of bodies" (ibid., p. 184).
3. *Values*: attitudes "more widely shared among different communities" (ibid., p. 184) than the first two components.
4. *Exemplars*: such as "concrete problem-solutions that students encounter from the start of their scientific education" (ibid., p. 187): in other words, textbook or laboratory examples.

Masterman (1970, p. 65) ordered these components by three types of paradigms:

(a) Metaphysical or meta-paradigms (refers to 2),
(b) Sociological paradigms (refers to 3), and
(c) Artefact or constructed paradigms (refers to 1 and 4).

Each paradigm shapes a disciplinary matrix according to which new knowledge can be structured, legitimized, and imbedded into the discipline's body of knowledge. Referring to Masterman, Burscheid used these types of paradigms to identify the scientific state of mathematics education in the development of four stages of a scientific discipline (see Burscheid 1983, pp. 224–227):

1. Non-paradigmatic science,
2. Multi-paradigmatic science,
3. Dual-paradigmatic science, and
4. Mature or mono-paradigmatic science (ibid., p. 224, translated[2]).

In the first stage, scientists originate the science by identifying its problems, establishing typical solutions, and developing methods to be used. In this stage, scientists struggle with the discipline's basic assumptions and a kernel of ideas; for instance, methodological questions of how validity can be justified and which thought models are relevant. In this stage, paradigms begin to develop, resulting in the building of scientific schools and shaping a multi-paradigm discipline. The schools' specific paradigms unfold locally within the single scientific group but do

[2]Any translation within this article has been conducted by the authors unless stated otherwise.

not affect the discipline as a whole. In stage three, mature paradigms compete to gain scientific hegemony in the field (Burscheid 1983, p. 226). The final stage is that of a mature scientific discipline in which the whole community shares more or less the same paradigm (ibid., p. 226).

Following the disciplinary matrix, Burscheid (pp. 226–236) identified paradigms in mathematics education and features at that time, according to which different scientific schools emerged and could be distinguished from one another, e.g., according to forms, levels, and types of schools, or according to reference disciplines such as mathematics, psychology, pedagogy, and sociology. The *constructed paradigms* dealt in principle with establishing adequate theories in a discipline. Concerning building theories, however, the transfer of the model of Masterman and Kuhn was difficult to achieve because symbolic generalizations and/or scientific laws can be built more easily in the natural sciences than in mathematics education. This is because mathematics education is concerned with human beings who are able to creatively decide and act in the teaching and learning processes. Burscheid doubted that a general theory such as those in physics could ever be developed in mathematics education (ibid., p. 233). However, his considerations led to the conclusion that "there are single groups in the scientific community of mathematics education which are determined by a disciplinary matrix…. That means that mathematics education is [still] heading to a multi-paradigm science" (ibid., p. 234, translated).

Burscheid's analysis was immediately criticized from two perspectives. Fischer (1983)[3] claimed that pitting mathematics education against the scientific development of natural science is almost absurd because mathematics education has to do with human beings (ibid., p. 241). In his view, "theory deficit" (ibid., p. 242, translated) should not be regarded as a shortcoming but as a chance for all people involved in education to emancipate themselves. The lack of impact on practice should not be overcome by top-down measures from the outside but by involving mathematics teachers bottom-up to develop their lessons linked to the development of their personality and their schools (ibid., p. 242). Fischer did not criticize Burscheid's analysis per se, but rather the application of a model postulating that all sciences must develop in the same way as the natural sciences towards a unifying paradigm (Fischer 1983).

Steiner (1983) also criticized the use of the models developed by Kuhn and Masterman. He considered them to be not applicable to mathematics education in principle, claiming that even for physics these models do not address specific domains in suitable ways, and in his view domain specificity is in the core of mathematics education (ibid., p. 246). Even more than Fischer, Steiner doubted that mathematics education would develop towards a unifying single-paradigm science. According to him, mathematics education has many facets and a systemic character

[3]Fischer also feared that if mathematics education developed towards a unifying paradigm, the field would be more concerned with its own problems, as was the case with physics, and, finally, would develop with its issues separated from societal concerns.

with a responsibility to society. It is deeply connected to other disciplines and, in contrast to physics, mathematics education must be thought of as being interdisciplinary at its core. The scientific development of mathematics education should not rely upon external categories of description and acceptance standards, but should develop such categories itself (ibid., pp. 246–247), and, moreover, it should consider the relation between theory and practice (ibid., p. 248).

Exactly such an analysis from the inside was proposed by Bigalke (1984) one year later. He analysed the development of mathematics education as a scientific discipline as well, but this time without using an external developmental model. He proposed a "suitable theory concept" (ibid., p. 133, translated) for mathematics education on the basis of nine theses. Bigalke urged a theoretical discussion and reflection on epistemological issues of theory development. Mathematics education should establish the principles and heuristics of its practice, specifically of its research practice and theory development, on its own terms. Bigalke specifically regarded it as a science that is committed to mathematics as a core area with relations to other disciplines. He claimed that its scientific principles should be created by "philosophical and theoretical reflections from tacit agreements about the purpose, aims, and the style of learning mathematics as well as the problematisation of its pre-requisites" (ibid., p. 142, translated).

Such principles are deeply intertwined with research programs and their theorizing processes. Many examples taken from the German didactics of mathematics were used to substantiate that Sneed's and Stegmüller's understanding of theory (see Jahnke 1978, pp. 70–90) fits mathematics education much better than the restrictive notion of theory according to Masterman and Kuhn, specifically when theories are regarded to inform practice. Bigalke (1984) described this theory concept in the following way:

> A theory in mathematics education is a structured entity shaped by propositions, values and norms about learning mathematics. It consists of a kernel, which encompasses the unimpeachable foundations and norms of the theory, and an empirical component which contains all possible expansions of the kernel and all intended applications that arise from the kernel and its expansions. This understanding of theory fosters scientific insight and scientific practice in the area of mathematics education. (p. 152, translated)

Bigalke himself pointed out that this understanding of theory allows many theories to exist side by side. It was clear to him that no collection of scientific principles for mathematics education would result in a "canon" agreed to across the whole scientific community. On the contrary, he considered a certain degree of pluralism and diversity of principles and theories to be desirable or even necessary (ibid., p. 142). Bigalke regarded theories as the link to the practice of teaching and learning of mathematics as well as being inspired by this practice, founding mathematics education as a scientific discipline in which theories may prove themselves successful in research and practice (Bigalke 1984).

2.2 Theories of Mathematics Education (TME): A Program for Developing Mathematics Education as a Scientific Discipline

Out of the previous presentation arose the result that the development of theories in mathematics education cannot be cut off from clarifying the notion of theory and its epistemological ground related to the scientific foundation of the field. Steiner (1983) construed this kind of self-reflection as a genuine task in any scientific discipline (see Steiner 1986) when he addressed the comprehensive task of founding and further developing mathematics education as a scientific discipline (see Steiner 1987c). At a post-conference meeting of ICME5 in Adelaide in 1984, the first of five conferences on the topic "Theories of Mathematics Education" (TME) took place (Steiner et al. 1984; Steiner 1985, 1986). This topic is a developmental program consisting of three partly overlapping components[4]:

- *Development of the dynamic regulating role* of mathematics education as a discipline with respect to the *theory-practice interplay* and *interdisciplinary cooperation*.
- Development of a *comprehensive view* of mathematics education comprising research, development, and practice by means of a *systems approach*.
- *Meta-research* and development of *meta-knowledge* with respect to mathematics education as a discipline (emphasis in the original; Steiner 1985, p. 16).

Steiner characterized mathematics education as a complex referential system in relation to the aim of implementing and optimizing teaching and learning of mathematics in different social contexts (ibid., p. 11). He proposed taking this view as "a meta-paradigm for the field" (ibid., p. 11; Steiner 1987a, p. 46), addressing the necessity of "meta-research in the field." According to Steiner, the field's inherent complexity evokes reduction of its complexity in favour of focusing on specific aspects, such as curriculum development, classroom interaction, or content analysis. According to Steiner, this complexity also creates a differential classification of mathematics education as a "field of mathematics, as a special branch of epistemology, as an engineering science, as a sub-domain of pedagogy or general didactics, as a social science, as a borderline science, as an applied science, as a

[4]This program was later reformulated by Steiner (1987a, p. 46):

– Identification and elaboration of *basic problems* in the orientation, foundation, methodology, and organization of mathematics education as a discipline

– The development of a *comprehensive approach* to mathematics education in its totality when viewed as an *interactive system* comprising research, development, and practice

– *Self-referent research* and *meta-research* related to mathematics education that provides *information* about the state of the art—the situation, problems, and needs of the discipline-while respecting national and regional differences.

foundational science, etc." (Steiner 1985, p. 11). Steiner required clarification of the relations among all these views, including the principle of complementarity on all layers, which means considering research and meta-research, concepts as objects and concepts as tools (Steiner 1987a, p. 48, 1985, p. 15). He proposed understanding mathematics education as a human activity; hence, he added an activity theory view to organize and order the field (Steiner 1985, p. 15). The interesting point here is that Steiner implicitly adopted a specific theoretical view *of the field* but points to the multiple perspectives *in the field*, which should be acknowledged as its interdisciplinary core.

Steiner (1985) emphasized the need for the field to become aware of its own processes of development of theories and models and investigate its means, representations, and instruments. Epistemological considerations seemed important for him, specifically concerning the role of theory and its application. In line with Bigalke, he proposed considering Sneed's view on theory as suitable for mathematics education, since it encompasses a kernel of theory and an area of intended applications to conceptualize applicability being a part of the very nature of theories in mathematics education (ibid., p. 12).

In the first TME conference, theory was an important topic, specifically the distinction between so-called *borrowed* and *home-grown* theories. Steiner's complementary view made him point to the danger of one-sidedness. In his view, so-called borrowed theories are not just transferred and used but rather adapted to the needs of mathematics education and its specific contexts. Home-grown theories are able to address domain-specific needs but are subjected to the difficulty of establishing suitable research methodologies on their own authority. The interdisciplinary nature of mathematics education requires regulation among the perspectives but also regulation of the balance between home-grown and borrowed theories (Steiner 1985; Steiner et al. 1984).

So, what is Steiner's specific contribution to the discussion of theories and theory development? Like other colleagues, such as Bigalke, he has pointed to the role of theories as being in the core of mathematics education as a scientific discipline, and he proposed the notion of theory developed by Sneed and Stegmüller (see Jahnke 1978, pp. 70–90) as being suitable for such an applied science. Steiner proposed complementarity to be a guiding principle for the scientific field and required investigating what complementarity means in each case of the field's topics. In this respect, the dialectic between borrowed theories and home-grown theories is an integral part of the field that allows the discipline to develop from its core and to be challenged from its periphery. In addition, Steiner emphasized that mathematics education as a system should reflect about its own epistemological basis, its own theory concepts and theory development, the relation between theory and practice, and the interrelation among all its perspectives. He has added that the specific view of mathematics education always incorporates some epistemological model of how mathematics and teaching and learning of mathematics are understood and that this is especially relevant for theories in mathematics education.

2.3 Post-TME Period

In the following decade, from 1992 up to the beginning of the 21st century, the discussion on theory concepts died down in the German community of mathematics educators while the theoretical diversity in the field grew. Considering the two main scientific journals, we identified scientific contributions from several theoretical communities addressing three topics related to the TME program (without any claim of completeness):

1. *Methodology*: *methodological* and thus *theoretical* aspects in interpretative research (Beck and Jungwirth 1999), *interviews* in empirical research (Beck and Maier 1993), *multi-methods* (Wellenreuther 1997); *explaining* in research (Maier 1998), *methodological* considerations on TIMSS (Knoche and Lind 2000);
2. *Methods in empirical research*: e.g., two special issues of *ZDM* in 2003 edited by Kaiser presented a number of methodical frameworks; and
3. *Issues on meta-research* about what mathematics education is, can, and should include: considerations on *paradigms* and the *notion of theory* in interpretative research (Maier and Beck 2001), *comparison research* (Kaiser 2000; Maier and Steinbring 1998; Brandt and Krummheuer 2000; Jungwirth 1994), and mathematics education as *design science* (Wittmann 1995) and as a *text science* (Beck and Maier 1994).

This short list indicates that—at that time—distinct theoretical communities seemed to share the need for methodological and meta-theoretical reflection. However, the German community of mathematics education as a whole did not—and still does not—share a common paradigm. In order to provide deeper insight into theory strands of German-speaking countries, two examples are presented.

The first one is the theory of learning activity that originates in activity theory developed by Joachim Lompscher. It is used today in several educational subjects: for example, Bruder has further developed and adapted this concept to the needs of mathematics education, and she and Schmitt will present this theory strand. The second theory strand is a specific view on semiotics presented by Dörfler and contrasted with Otte's view on signs as a vehicle for doing mathematics as a human activity.

The theory of learning activity provides a general educational theory that has been borrowed then applied and adapted to mathematics education, while Dörfler bases his work profoundly in the philosophies of Peirce and Wittgenstein and reconstructs mathematics as a kind of game using diagrams in a more home-grown way.

Chapter 3
Joachim Lompscher and His Activity Theory Approach Focusing on the Concept of Learning Activity and How It Influences Contemporary Research in Germany

Regina Bruder and Oliver Schmitt

3.1 Introduction

The concept of activity is a psychological construct that connects man and his development to culture and society. This concept was shaped substantially by Vygotsky, Leontiev, and Luria and developed further in the German-speaking countries by Lompscher[1] in particular. The activity theory, which follows this line of tradition, has often been assigned to social constructivist approaches (Giest and Lompscher 2006, p. 231; Woolfolk 2008, p. 421). Lompscher elaborated the concept of learning activity with regard to teaching practice and applied it to several

[1]Joachim Lompscher (1932–2005) is considered the "founder of educational psychology and of the psychology of learning activities in the GDR" (Rückriem and Giest 2006, p. 161, translated). Focal points of his academic work were the development of mental abilities, the training of learning activities, the cultural-historical school of Soviet psychology and the associated activity theory, and aspects of its development in the history of psychology. He studied psychology and education in Moscow and defended his doctoral thesis in Leningrad in 1958 on the subject "On the understanding of children of some spatial relationships" (translated). He subsequently worked at the Humboldt University of Berlin, moved to the German Central Institute of Psychology (DZPI) in 1962, and from 1966 was there in a leading position for practical teaching projects and issues in the mental development of children. He habilitated in Leipzig in 1970 and was subsequently appointed Professor of Educational Psychology at the Academy of Educational Sciences (APW) in Berlin. After German reunification in 1991, he worked at the Institute of Learning and Teaching Research at the University of Potsdam (For an obituary and bibliography, see Rückriem and Giest 2006).

R. Bruder (✉) · O. Schmitt
Fachbereich Mathematik, Technical University Darmstadt, Schlossgartenstr. 7, D-64289 Darmstadt, Hessen, Germany
e-mail: bruder@mathematik.tu-darmstadt.de

O. Schmitt
e-mail: oschmitt@mathematik.tu-darmstadt.de

A. Bikner-Ahsbahs et al., *Theories in and of Mathematics Education*, ICME-13 Topical Surveys, DOI 10.1007/978-3-319-42589-4_3

subjects. The core objective of teaching is the training of learning activity, which is aimed at acquiring social knowledge and competence and requires specific means under specially arranged conditions. The concepts of learning tasks and orientation bases of learning actions are closely linked to the concept of learning activity. These conceptual bases are briefly presented in Chap. 2, whilst Chap. 3 refers to current applications of the activity theory in German-speaking research on teaching methodologies.

Contemporary activity theory became an interdisciplinary discourse mainly through the works of Engeström in the field of the emerging labour research. This line of research sees itself as an "intervention approach to the study of changes and learning processes at work, in technology and organisations" (Engeström 2008, p. 17, translated) and is based on the tradition of the cultural-historical activity theory. In his theory and intervention methodology, Engeström dealt with the solution to practical social issues and, among other things, also provided valuable impulses for the development of teaching staff in schools (Engeström 2005).

Increased attention is also given to activity theory in international discussions on teaching methodologies (see Mason and Johnston-Wilder 2004), with the German-speaking countries contributing concepts such as describing the use of digital tools in mathematics classes (see Ladel and Kortenkamp 2013).

3.2 Conceptual Bases

The central concept of *activity* has been described as "the specifically human form of activity, of interaction with the world in which man changes it and himself at the same time" (Giest and Lompscher 2006, p. 27, translated). Activity takes place through the conscious influence of a subject on an object in order to shape the latter in accordance with the motive of the activity. To this end, such actions (material or spiritual) are performed within one activity line that each time realises certain sub-goals through to the ultimate product of the activity. At the same time, the concept of operation serves to further distinguish another form of subordinate activity that differs from actions by the fact that operations result from concrete conditions for action and pass in an automated manner without conscious control or goal formation. These represent shortened actions.

In the course of their lives, humans, in their confrontations with the world, develop various forms of activity, such as play, work, or learning activities that feature different characteristics in each case. For schools and for didactic research, the concept of learning activity has been of key importance. There, learning activity has been understood "as the activity aimed specifically at acquiring social knowledge and competence (learning topics) for which purpose specific means (learning resources) under specially arranged conditions have to be adopted." (Giest and Lompscher 2006, p. 67, translated). According to Lompscher (1985), three essential subjective requirements must be met on the part of the learners to achieve a learning activity:

- *Concrete learning goals as individual mental anticipation of the desired results and of the activity aimed at such results.*
- *Learning motives as the motivational basis to perform certain activities.*
- *Learning activities as*:

> Relatively closed and identifiable steps, structured in terms of time and logic, in the course of the learning activity, which realise a concrete learning goal, are driven by certain learning motives and are executed, according to concrete learning conditions, by the use of external and internalised learning resources in a specific sequence of sub-actions each time. (p. 46, translated)

The aim of school education has been without doubt to stimulate and promote learning activities in the learner. For instance, for mathematics classes, tasks have traditionally been perceived as a key creative resource of the teacher. Within the framework of the activity theory, suitable learning tasks have been understood as requests to perform learning actions (Bruder 2010, p. 115). There, a distinction has been made between the requirements imposed by teachers in relation to the learning topics and the learning tasks assigned by the learners to themselves. When planning classes, attention should be paid to allow as much scope as possible "for the construction of individually suitable learning tasks" (Bruder 2008, p. 52, translated).

Learning actions implemented in learning activity can be of a very different nature. According to Lompscher, various categories of learning actions can be distinguished depending on the learning task dominating in a given learning situation. These include, for instance (Lompscher 1985):

- observing objects, processes and situations according to pre-set or independently developed criteria;
- collecting, compiling, and processing data or materials for specific purposes and under certain aspects;
- performing actions of a practical or concrete nature to manufacture a product or to change it with regard to certain quality and effectiveness parameters;
- presenting circumstances orally and in writing for specific purposes whilst considering certain conditions;...
- assessing and evaluating third-party or own performance or behaviour or a given event with regard to certain measures of value;
- proving or refuting views in an arguing manner on the basis of certain positions, findings or facts;
- solving problems of various structures and contents; and
- practising certain actions (p. 48, translated).

These actions can be developed and recalled by learners in different ways (level of awareness and acquisition of an action). "One action can be performed at a level of relatively unfocused trial and error behaviour, whereas another one would proceed as a target-oriented search, adequate as per circumstances, with purposeful implementation of correlations recognised" (Lompscher 1985, p. 49, translated).

This issue can be described in a more differentiated way through an analysis of the structure of learning actions. Within an action, three different parts have been distinguished: the orientation part, the performance part, and the control part (see Giest and Lompscher 2006, p. 197). In the orientation part, an orientation basis is formed as a provisional idea of a task (Galperin 1967, p. 376) on the basis of which the action is eventually performed and the result of which is controlled with regard to previous goals. The concept of orientation basis was developed by the Soviet educationalist Galperin and extensively appreciated by didactic research in the GDR, particularly by Lompscher. According to Lompscher, the following issues in relation to requirements and the learning topic are relevant in the formation of the orientation basis (Giest and Lompscher 2006):

- What (requirement structure, sequence of sub-actions)
- How (examination conditions, resources, methods, quality of the action)
- Why (reason for the action, its inner connections)
- What for (classification of the action in overall connections, possible consequences, etc.) (p. 192, translated).

A distinction has basically been made between three different types of orientation (Giest and Lompscher 2006, pp. 192ff)—here reflecting the designations by Bruder (2005, p. 243):

- *Trial orientation* (*Probierorientierung*) designates an incomplete orientation basis entailing an action after trial and error; awareness of the procedure is very limited only and a transfer is hardly possible on that basis.
- In *pattern orientation* (*Musterorientierung*), some aspects and conditions of a requirement are recognised and associated with an example (pattern) already solved; the orientation basis is complete but transferable to a delimited area only, as no comprehension of the entire requirement class takes place.
- *Field orientation* (*Feldorientierung*) designates a complete general orientation basis resulting from an independent analysis of the requirements of a given field of knowledge or thematic field, which therefore allows for good transferability of the knowledge and actions acquired to new requirements.

If the requirement is, for instance, about solving a linear equation, learners with trial orientation would rather proceed by making transformations in an unsystematic manner or perhaps guess the figures and possibly even be successful. With pattern orientation they could also try to trace a systematic approach on the basis of an example already known to them, which would possibly allow for limited transferability to similar examples. Finally, in case of a developed field orientation, general strategies could be used, such as a separation between variable and constant terms on both sides of the equation.

By means of learning actions, depending on the arrangement of the learning environment, different orientation bases can be promoted in learners. Within the scope of practising processes during introductions to solving quadratic equations, the examining operation as to which type of equation is actually involved will become less important. Learners will be aware of what the current issue is about. Schematic practising can therefore only bring assurance and automatisms in processing algorithmic step sequences. Still, this does not lead to a transferable acquisition of the object. So, for instance, when solving a given quadratic equation within the scope of an aptitude test for vocational training, it will first have to be recognised that indeed such a type of equation is involved. If such an assignment is successful, the solution methods available will possibly be activated (development of example-based orientation). Such a task will only make higher demands on orientation building if the relevant equation type is still unknown or as part of mixed exercises at a later date.

If solution methods (graphic solutions, calculation formulae) can be activated at least at the level of example-based orientation, the relevant task can mostly be solved, except for some calculation or presentation errors. If such recognition of the equation type is not successful, various search processes are initiated, often with incorrect schema assignments, or the attempt at solution is discontinued altogether. In such a situation, intuitive reference is made to the basic concepts available and even to everyday experiences in the form of empirical generalisations. This, according to Nitsch (2015), would also explain, for instance, the differing stability of error patterns, whilst competing example-based orientations are available, partly incorrect or inadequate, which can be recalled depending on the context.

The approach of orientation bases yields important conclusions when considering a long-term development of fundamental mathematical competencies, such as in mathematical argumentation. To achieve high quality in the training for learning action "proving or refuting in an arguing manner" in mathematics classes, knowledge relevant to action is required. In particular, such knowledge is necessary as to which arguments are admissible in mathematics and which methods of conclusion are possible in order to be able to develop a field orientation for a processing strategy in relation to a given proof-related task. If such background knowledge is lacking, any transfer of this procedure, even with simple justifications (are all rectangles trapezia, too?), to other mathematical contents, such as proofs of divisibility, will hardly succeed. Instead, attempts are made to develop further example-based orientation within the new scope. Here, in schematic practising processes, the procedure is just transferred from one task to an analogous task, without awareness of what the procedure actually consists of. Such reflection processes with the building of knowledge are part of the training for a given learning action (in stages) and a necessary prerequisite for developing field orientation with the corresponding demands. If the demands remain at the level of analogous tasks, there will be no need to develop orientations of a higher quality and thus to advance the respective learning action.

In order to stimulate an orientation as far-reaching as possible at an early stage of the learning process, i.e., the formation of a learning goal, a teaching strategy, going back to Davydov (1990), of the rise from the abstract to the concrete has to be developed. As a first interim result in the learning process, a so-called starting abstract (*Ausgangsabstraktum*) is developed together with the learners, which maps, relates, and anchors the essential characteristics of the learning topic and offers a framework for the continuation of the teaching process. The starting abstract is thus “the result of learning activity already and as such the starting point for rising to the concrete” when further working with concrete contents (Giest and Lompscher 2006, p. 222, translated). Due to the heterogeneity of the learners, the tasks assigned by the teacher, which first have to be transformed into individual learning tasks, should allow for orientation at different levels to give the learners a chance to reach the individual zone of the next development stage in terms of Wygotskij (Bruder 2005, p. 243).

An approach to learning phenomena based on the activity theory by Lompscher includes the following aspects (Lompscher 1990):

- *the quality of the learning motives and goals at the activity level, which determine the concrete purpose and process of the learning actions;*
- *the interrelations between the activity and action (and also operation) levels, for instance, with regard to contradictions between activity motivation and concrete situational action motives; and*
- *the cognitive, metacognitive, emotional, motivational and volitive regulation bases, and the process structure of learning actions and learning outcomes (in terms of psychological changes).*
- *This and other questions can be worked on at different analysis levels, starting (1) with the most general components, relations and determinants of the macrostructure of the activity,* via *(2) an analysis of concrete classes of learning activities, such as learning from texts or solving problems with certain, although different, categories of learners, through to 3. the microanalysis of elementary components and processes based on performance of the action (p. 1f, translated).*

3.3 Exemplary Applications of the Activity Theory

Applications of the activity theory in German-speaking countries primarily refer to the analysis and formation of learning activities in connection with their corresponding knowledge, abilities, and skills. In parallel, various types of competence modelling on the basis of concepts of the activity theory have been performed or operationalised for diagnosis.

A consistent implementation of the activity theory according to Lompscher and in connection with Davydov was presented in the works on a theory of learning tasks by Dietz and associates (reported in Brückner 2008).

Mann (1990) explained learning how to read and write and do arithmetic on the basis of the activity theory and demonstrates how successful this approach has been for the development of learning surroundings even for people with intellectual disabilities.

The idea of the cognitive process as a unity of analysis and synthesis, going back to Rubinstein (1973), was expanded by Lompscher to describe the structure of mental abilities with the components *mental operations* and *process qualities*. The presentation by Lompscher (1975, p. 46) on the model interrelations between analytical and synthetic operations in mental activities was taken up by Bruder and Brückner (1989). According to this approach, identifying and realising mathematical contents can be described as elementary actions on the basis of defined mental operations. Empirical studies provide preliminary indications of evidence that these two elementary actions can be distinguished and also of basic actions of a more complex construction, such as describing and justifying each time in relation to given mathematical concepts, connections, or processes (see Nitsch 2015). Such a hierarchical approach to describing learning actions results in a heuristic construction for learning and test tasks (see the general approach to the task theory in Bruder 2003) which has already proven its worth in theoretical competence modelling. These action hierarchies are currently being used in a project aimed at describing the requirements for the central school-leaving examinations in Austria in a four-stage competence structure model for action dimensions in operating, modelling, and arguing (see Siller et al. 2015). Such a theoretical background was also used for the construction of items within the scope of the project HEUREKO on the empirical clarification of competence structures in a specific mathematical context, notably the changes of representation of functional relationships (see Nitsch et al. 2015).

Boehm (2013) used basic positions of the activity theory to establish curricular objectives for mathematical modelling at Secondary Level I. The theoretical framework for the analysis of modelling activities that he elaborated allows for a differentiated model description of the action elements in mathematical modelling. This also includes the successful involvement and clarification of problem-solving activities in modelling.

Mathematical problem-solving competence can be interpreted, from an activity theory angle, as variously pronounced mental agility where mental agility represents a marked process quality of thinking [see the construct of process qualities in Lompscher (1976)]. According to Lompscher (1972, p. 36), content and the progress of learning actions are decisive for the result. Bruder's (2000) operating principle in acquiring problem-solving competence was that through the acquisition of knowledge about heuristic strategies and principles, insufficient mental agility can partly be compensated. This approach was transferred to a teaching concept about learning how to solve problems in four stages building on each other, and the corresponding effects at student level have been empirically proven (Bruder and Collet 2009; Collet and Bruder 2008).

Nitsch (2015) investigated typical difficulties of learning in changes of representation of functional relationships and interpreted these as incomplete orientation bases. Existing error patterns could be described as inadequate patterns. In this way, and in connection with the concept of basic ideas (Vom Hofe 1995), a tentative explanation is provided about mechanisms to activate certain mathematical (error-) ideas.

In terms of orientation bases, there was a discussion in the 1970s both in the GDR and in a Western response by critical psychology about whether another type going beyond the field orientation should be added to the previously mentioned orientation types. The intention of this orientation type was to describe the creative handling of open issues that did not already have any known or generally recognised solutions at hand. Taking up this discussion and providing a response to the teaching strategy of the rise from the abstract to the concrete, Schmitt (2013) developed a concept to promote reflective knowledge (Fischer 2001; Skovsmose 1989) in mathematics classes in a targeted manner.

Feldt (2013) uses concepts of the activity theory as a background to conceptualise minimum standards. The activity theory offers opportunities to operationalise learning goals through its central concepts of learning action and learning task but also through the construct of the acquisition quality of knowledge (see Pippig 1985) in connection with the orientation bases of the learning actions. In particular, the quality feature of availability of knowledge, which has been highly relevant in conceptualising minimum standards, is being discussed with a view to a possible gradation in the style of Sill and Sikora (2007) and is being further refined with regard to such gradation.

Chapter 4
Signs and Their Use: Peirce and Wittgenstein

Willi Dörfler

4.1 Introductory Remarks

It is obvious that mathematicians throughout history have used signs of various kinds, such as symbols, diagrams, graphs, and formulae, but they also occur in everyday language and scientific language. The technical symbols and formulas of mathematics have contributed in particular to its specific status, and many learning difficulties have been attributed to these characteristics, which are viewed as turning mathematics into a highly abstract and inaccessible field of scientific enquiry. Even for the most basic mathematical activities such as arithmetic calculations, the use of number symbols is unavoidable, and it can also be said that much of the strength and relevance of mathematics for applications derives from its symbolic techniques. One could express this by stating that the "formula" was one of the great cultural inventions and intellectual innovations, comparable in its ramifications to those of the wheel. The use of symbolic techniques within mathematics, such as in proofs, needs no further discussion. It is also virtually impossible to translate mathematics into any kind of vernacular, and most mathematical "narratives" are rather misleading or missing the mathematical point. To understand mathematics, one has to do it; this doing in a very deep sense is an activity with signs and based on signs, as should become even clearer from the following considerations.

So far most people concerned in some way or the other with mathematics will agree with what was stated above. Pronounced differences show up when one turns to what one can term the meaning of the signs and symbols of mathematics. In the common understanding, signs are used to designate something that is different from and independent of the sign, namely, the object of the sign, and this object is viewed as the source of the meaning of the sign. Often the signs are considered as

W. Dörfler (✉)
Institut für Didaktik der Mathematik, Alpen-Adria-Universität Klagenfurt,
Universitätsstraße 65, 9020 Klagenfurt, Austria
e-mail: willi.doerfler@aau.at

A. Bikner-Ahsbahs et al., *Theories in and of Mathematics Education*,
ICME-13 Topical Surveys, DOI 10.1007/978-3-319-42589-4_4

being secondary to what they designate and arbitrary and neutral with respect to the mathematical content. Their main use in this view is to communicate and express the mathematical ideas. Hersh (1986, p. 19), for instance, compared mathematics with music, where according to him the score has solely the role of noting the music which is already there before the score. The signs and notations in this view have no influence on invention and creation in mathematics or music. An extreme position in this vein was taken by Brouwer see Shapiro (2000), who considered mathematics to be a purely mental "construction" not dependent on any sign system. In a general way, in all these positions of mathematical realism mathematical signs and notations have been viewed as describing what have been termed mathematical objects, whatever those might be and wherever they might be located. Thus numerals denote numbers and diagrams denote geometric objects. Only algebraic formulas have sometimes been spared this descriptive role, yet they have then been reduced to a purely technical means for calculations and proofs. I will not continue these ontological and philosophical issues any further, but these short hints should serve to make the possible impact of the views taken by Peirce and Wittgenstein more conspicuous.

4.2 Charles Sanders Peirce

Peirce (1839–1914) was an American mathematician, logician, and philosopher. From among his comprehensive works, only his fundamental work in semiotics can very briefly be considered here. Peirce developed a complex and comprehensive theory of signs by devising a multilevel categorization of signs, starting with the differentiation into index, icon, and symbol. With Peirce, the sign in itself has a triadic structure of "object-representamen-interpretant," but we will not go into any details here. Interestingly, for decades mathematics educators apparently have not taken note of the potential of the theories presented by Peirce. Yet Peirce was interested in educational questions and has written a very interesting draft for a textbook on elementary arithmetic [see the two articles by Radu in Hoffmann (2003)].

To the best of my knowledge, it was due to the initiative taken by Michael Otte in some of his papers (see Otte 1997, 2011) that the relevance of the semiotics of Peirce was recognized by a growing number of mathematics educators in Germany and elsewhere. It is impossible to adequately present the work by Otte with regard to Peirce here because it is very complex and comprehensive. He puts Peirce and his semiotics into the context of philosophy, epistemology and ontology by relating it to many other strands of thought in this realm but pays less attention to the concrete mathematical activities on and with signs. Rather, the papers by Otte furnish a powerful background and basis for more detailed investigations into/about how and which signs are used in mathematics and especially in mathematics learning. On the other hand, his papers show and explicate deliberations in Peirce that may be more general and fundamental. But it is also sensible to investigate—as it will be done here—a Peircean notion, such as diagrammatic reasoning, independently from other dimensions of Peircean semiotics and its philosophical ramifications. In a pointed way, one could say that in Michael Otte the purview of a sign is the whole of life, experience, and cognition, whereas here we focus on its important role in doing and learning mathematics. To give the reader a flavour of the work by Otte, it is instructive to cite from the abstracts of Otte (1997; my translation from the German original) and of Otte (2011):

> Peirce treats the concepts meaning, (natural) law, continuum—and some others like representation or mind—as synonyms. By that they all acquire those paradoxical qualities which have been since long discussed for the example of the continuum and which recently have been addressed in different contexts, as in systems theory. The meaning of a sign, for example, for sure cannot be separated from its application—what is already stipulated by the Pragmatic Maxim of Peirce. On the other hand, it cannot be identified either with a single application or with some well-defined set of applications but it rather rests on the general conditions for possible applications. The notion of sign and the concept of the continuum are the two pillars on which Peirce's phenomenological epistemology is based. The latter shall be elucidated first, through the relation to the history of mathematics; and second, through the comparison with other phenomenological positions during the foundational crisis of mathematics. The significance of mathematics results from the fact that in mathematics, the two pillars mentioned most deeply confront each other. (Otte 1997, p. 175)

> One of the most salient arguments in favour of a semiotic approach... claims that semiotics is most appropriate for treating the interaction between socio-cultural and objective aspects of knowledge problems. If we want to take such claims seriously, however, we have to revise our basic conceptions about reality, existence, cognition, and cultural development. The semiotic evolutionary realism of Charles S. Peirce provides—or appears to provide—an appropriate basis for such intentions. Man is a sign, Peirce famously said, and "thought is more without us than within. It is we that are in it, rather than it in any of us" (Peirce CP 8.256). As there is no thought without a sign, we have to accept thoughts, concepts, theories, or works of art as realities *sui generis*. Concepts or theories have to be recognized as real before we ask for their meaning or relevance. (Otte 2011, p. 313)

An important early contribution to the dissemination of the semiotics of Peirce was Hoffmann (Hoffmann 2005a), which explicated many aspects of Peircean semiotics, especially with emphasis on mathematics. A related work is that by Stjernfelt (2000), which also contains very worthwhile interpretations of ideas and notions in Peirce. Much of this work was concerned with Peirce's general sign theory and its philosophical dimensions. Within mathematics education, in addition to the triadic structure of sign, the notion of diagram and diagrammatic thinking was mainly exploited. It should be noted that for Peirce, signs always possessed an object that they explored in an ongoing semiotic process; the only exception was diagrams, for which Peirce allowed the object to be fictional or ideal, especially with respect to mathematics. Before concentrating on the concept of diagrammatic thinking, which appears to be of special value for mathematics and the learning of mathematics, some more references on the work on (Peircean) semiotics within German mathematics education are included: Hoffmann (2003, 2005), Hoffmann et al. (2005), Kadunz (2010, 2015). Of course, on an international level semiotics in general and Peircean notions in particular have also received growing attention. Some publications illustrate this ever-extending tendency: Rotman (2000); the contributions to special issues in the journals *ESM*, *ZDM*, and JMD by authors including Presmeg, Saenz-Ludlow, and Radford; and Radford et al. (2008). In addition to the publications that have an explicit focus on semiotics, one could refer to the vast literature on visualization and representation. Yet because in these the signs have mostly been considered in their descriptive and representational function (see below), this is beyond the scope of this contribution. In addition to Peirce, there have been other semiotic traditions and theories which have been exploited in mathematics education; for instance, Duval (1995). We now turn to the notion of diagram and diagrammatic thinking in the form of a liberal interpretation of the ideas of Peirce based on Dörfler (2004, 2006, 2008), where one can find a host of examples for diagrams and diagrammatic reasoning.

4.3 Diagrams and Diagrammatic Thinking

Peirce (3.363 in Collected Papers, this means paragraph 363 in Volume 3 according to the standard way of citing from the papers by Peirce) made the following comment, among others, on a basic feature of mathematics:

> It has long been a puzzle how it could be that, on the one hand, mathematics is purely deductive in its nature, and draws its conclusions apodictically, while on the other hand, it presents as rich and apparently unending a series of surprising discoveries as any observational science. Various have been the attempts to solve the paradox by breaking down one or other of these assertions, but without success. The truth, however, appears to be that all deductive reasoning, even simple syllogism, involves an element of observation; namely, deduction consists in constructing an icon or diagram the relations of whose parts shall present a complete analogy with those of the parts of the object of reasoning, of experimenting upon this image in the imagination, and of observing the result so as to discover unnoticed and hidden relations among the parts.... As for algebra, the very idea of the art is that it presents formulae, which can be manipulated and that by observing the effects of such manipulation we find properties not to be otherwise discerned. In such manipulation, we are guided by previous discoveries, which are embodied in general formulae. These are patterns, which we have the right to imitate in our procedure, and are the icons par excellence of algebra.

I have chosen to stick to the term *diagram* as it has been used by Peirce and others, though I am aware that this term might cause some misunderstandings and arouse false expectations. First of all, the reader should dismiss all geometric connotations. This can be seen from the above reference to Peirce, who includes formulas of all kinds in his notion of diagram (or icon). What is important are the spatial structure of a diagram, the spatial relationships of its parts to one another, and the operations and transformations of and with the diagrams. The constitutive parts of a diagram can be any kind of inscriptions such including letters, numerals, special signs, or geometric figures.

Peirce did not nor will I give a general definition of the notion of diagram. Instead, several descriptive features of diagrams are presented. Diagrams are based on a kind of permanent inscription (paper, sand, screen, etc.). Those inscriptions are mostly planar, but some are 3-dimensional, such as models of geometric solids or manipulatives in school mathematics. Mathematics at all levels abounds with such inscriptions: number lines, Venn diagrams, geometric figures, Cartesian graphs, point-line graphs, arrow diagrams (mappings), arrows in the Gaussian plane or as vectors, and commutative diagrams (category theory); but there are also inscriptions with a less geometric flavour: arithmetic or algebraic terms, function terms, fractions, decimal fractions, algebraic formulas, polynomials, matrices, systems of linear equations, continued fractions, and many more. There are features common to some of these inscriptions that contribute to their diagrammatic quality as it is understood here. However, I emphasize that not every inscription that occurs in mathematical reasoning, learning, or teaching has a diagrammatic quality. Quite a few of what are taken as visualizations or representations of mathematical notions and ideas do not qualify as diagrams since they lack some of the essential features. This is mostly the precise operative structure that for genuine diagrams permits and invites their investigation and exploration as mathematical objects. Some widely shared qualities of diagrams, or rather of inscriptions when used as diagrams, are proposed in the following:

- Diagrammatic inscriptions have a structure consisting of a specific spatial arrangement of and spatial relationships among their parts and elements. This structure often has a conventional character.
- Based on this diagrammatic structure, there are rule-governed operations on and with the inscriptions by transforming, composing, decomposing, or combining them (calculations in arithmetic and algebra, constructions in geometry, and derivations in formal logic). These operations and transformations could be called the internal meaning of the respective diagram (compare to Wittgenstein on meaning). Depending on the operations and transformations applied, an inscription might give rise to essentially different diagrams. Thus, a triangular inscription will be a general or isosceles triangle, depending on which of those properties is used in diagrammatic arguments; this is similar to the same card playing different roles in different card games.
- Another set of conventionalized rules governs the application and interpretation of the diagram within and outside of mathematics, i.e., what the diagram can be taken to denote or model. These rules could be termed the external or referential meaning (algebraic terms standing for calculations with numbers or a graph depicting a network or a social structure). The two meanings closely inform and depend on each other.
- Diagrammatic inscriptions express (or can be viewed as expressing) relationships by their very structure, from which those relationships must be inferred based on the given operation rules. Diagrams are not to be understood in a figurative but rather in a relational sense (such as a circle expressing the relation of its peripheral points to the midpoint).
- There is a type-token relationship between the individual and specific material inscription and the diagram of which it is an instance (such as between a written letter and the letter as such).

- Operations with diagrammatic inscriptions are based on the perceptive activity of the individual (such as pattern recognition) that turns mathematics into a perceptive and material activity.
- Diagrammatic reasoning is a rule-based but inventive and constructive manipulation of diagrams for investigating their properties and relationships.
- Diagrammatic reasoning is not mechanistic or purely algorithmic; it is imaginative and creative. Analogy: the music of Bach is based on strict rules of counterpoint but is highly creative and variegated.
- Many steps and arguments of diagrammatic reasoning have no referential meaning, nor do they need any.
- In diagrammatic reasoning the focus is on the diagrammatic inscriptions irrespective of what their referential meaning might be. The objects of diagrammatic reasoning are the diagrams themselves and their established properties.
- Diagrammatic inscriptions arise from many sources and for many purposes: as models of structures and processes, by deliberate design and construction, by idealization and abstraction from experiential reality, etc., and they are used accordingly for many purposes.
- Efficient and successful diagrammatic reasoning presupposes intensive and extensive experience with manipulating diagrams. A comprehensive "inventory" of diagrams, their properties, and relationships supports and facilitates the creative and inventive usage of diagrams. An analogy: expert chess players have command over a great supply of chess diagrams that guide their strategic problem solving. Consequence: learning mathematics has to comprise diagrammatic knowledge of a great variety.
- Diagrams can be viewed as ideograms, such as those in Chinese writing systems. They are not translations from any natural language or abbreviations of names and definitions; by their diagrammatic structure, they "directly" present (to the initiated user!) their intended meaning. The latter usually is a system of relationships (between the elements or the parts of the diagram) and of operations and transformations.
- Diagrams are composed of signs of different characters in the sense of Peirce. There are icons, indices, and symbols as well, and a whole diagram has iconic and symbolic functionality if in itself it is considered to be a sign in the sense of Peirce.

To be understood and used appropriately, diagrams need to be described in natural language and specific terms relating to the diagram. These descriptions and explanations cannot be substituted for the diagram and its various uses, however. In relation to the diagram and its intended relations and operations, this is a meta-language about the diagrams, which also focuses attention and interest on its relevant aspects and activities. It is similar to the way in which the legend on a map of a city explains how to use that map appropriately. Generally, diagrams are imbedded in a complex context and discourse, which is better viewed as a social practice.

- Diagrams are extra-linguistic signs. One cannot speak the diagram, but one can speak about the diagram. In this sense, diagrams are irreducible entities of mathematics (there is no mathematics without "formulas"), yet their properties can be named by words and formulated as theorems. Thus, on the other hand (specialized) language (as extension of natural language) is equally indispensable.

As a final remark: it would be misleading to consider diagrams as mathematical objects. They are the objects and the means of mathematical activity for which we do not have to view them as designating mathematical objects. This emphasis on activity and concrete operations with signs leads us to Wittgenstein's views.

4.4 Wittgenstein: Meaning as Use

The Austrian philosopher Ludwig Wittgenstein (1889–1951) dedicated a great part of his work to the philosophy of mathematics (e.g., Wittgenstein 1999), proposing radically alternative views on the basic character of mathematics. Together with other features of his writings, this might have prevented any notable recognition within mathematics education. Therefore, this contribution will (also) try to alert the community of mathematics education to the potential of the ideas of Wittgenstein which might (also) influence general attitudes and basic orientations of the concrete teaching in the classroom. A caveat is, of course, that only a few aspects can be treated here and these in only a rather superficial way. The interested reader is referred to Dörfler (2013a, 2014) and the vast literature on Wittgenstein's philosophy of mathematics, for instance, Kienzler (1997) or Mühlhölzer (2010).

Contrary to the traditional view, Wittgenstein views the meaning of many signs, words, and symbols in general and of mathematics as well to reside in the use made of those signs in what he calls language games or sign games. Thus, signs do not express a meaning that exists independently of the sign game and that is given by something outside of the sign game that the signs refer to and denote. For mathematics, then, the meaning of the signs, symbols, and diagrams does not come from outside of mathematics but is created by a great variety of activities with the signs within mathematics. This resonates very closely with the diagrammatic reasoning described above (though Peirce would hold that thereby some independent "object" is investigated, contrary to the position taken by Wittgenstein, which is strongly non-metaphysical and anti-platonistic). Wittgenstein introduces the metaphor of mathematics as a game, in particular by pointing to chess. In chess, the figures receive all their meaning from the rules of the game, and they do not refer to anything outside of the system of rules. The figures correspond to the signs in mathematics and the game rules correspond to the rules in mathematics for calculating, manipulating, and deriving (i.e., the diagrammatic rules in the above sense). This game metaphor helps to solve many puzzles in math: Consider the "number" zero. There has been and continues to be a great deal of discussion about what this sign denotes and how it could designate a number. In the Wittgensteinian sense, the meaning of "0" is determined and presented by the rules for how we calculate with it; $5 + 0 = 5$ or $0 \times 6 = 0$, for example, reflect the origin of zero from the place value systems. Thus, there is no mystery and no miracle about zero if you do not ask questions that are outside the purview of math (what Wittgenstein in a telling way calls the prose of mathematics). Very similar considerations apply to the empty set, the "number" -1 and first and foremost to i, the imaginary unit which simply is determined by the rule that $i \times i = -1$. It is a very helpful and sober way of thinking in this way to consider the respective number systems as number games where the meaning of the number signs flows from how they are calculated and not from a mystical reference, say, to "nothing" or negative or imaginary magnitudes. About those mathematical entities we can only know what is shown to us by the results of the calculations within the number games.

To pursue this line of thinking further, we turn to the notion of grammar and grammatical proposition as used by Wittgenstein. He says that mathematical propositions do not describe factual situations as do propositions in science because there are no independent mathematical objects those propositions could be about. In his view, mathematical propositions are instead rules for how to use the terms and signs involved in their formulation as they are developed within the various sign games of mathematics. Or to put in still another way, in mathematics the propositions are used as rules, e.g., in proofs and calculations, though in mathematical prose they are interpreted as accounts of mathematical facts in a mathematical world. Examples for nonmathematical grammatical propositions would be: "White is brighter than black," "Every rod has a length," "Nothing can be red and blue at the same place," or "Every finite set has a number." Every arithmetic "fact" in this view is just another rule and not the description of an eternal and absolute truth about numbers. The concerns of many philosophers and sociologists about the status of, say, "$2 + 2 = 4$" dissolve when one takes this as a rule, which of course then can neither be verified nor falsified in an empirical interpretation. For Wittgenstein, the whole notion of truth against this background makes no sense since rules are neither true nor false. Rules have to be accepted; they require consent, which is often motivated by a kind of practicability and viability. Rules are outside of all aspects of time (in addition to questions such as when they were established or abolished) or at least this is the way we use rules. Think again of chess as a metaphor for the sign games of mathematics. The rules of chess usually are not viewed as being true or eternal, and one can refuse to accept them but then one will not be playing chess anymore. Such a view fundamentally changes one's attitudes and relations to mathematics and the learning of mathematics. The practice and fluency in sign games is now the centrepiece and not the mental grasp of ideal and abstract objects or of "ideas" which are just denoted and represented by the mathematical signs. The learner has to indulge in the mathematical "games" whereby meaning and understanding gradually will develop. In mathematics, meaning cannot be imported from outside but emerges inside it through manifold activities. Wittgenstein was often blamed for the apparent conventional and thus possibly arbitrary character of mathematics derived from his views. Yet to counter this, one can point to the fact that many basic rules (axioms) are motivated by practical or theoretical demands and that many other rules are then derived from given ones by proofs and calculations. On the other hand, there is in fact a great liberty regarding the rules according to which one wants to do the mathematics and this holds as well for the logic involved.

As Wittgenstein says, mathematics can be viewed as the grammar or the grammatical study of its signs and terms. This proves especially helpful wherever a notion of the "infinite" turns up, which notoriously poses great obstacles for learners. Historically it is interesting that Leibniz remarked that for his infinitesimals such as dx, one should not look for referents, that is, objects that are denoted by them. He took the view that they are completely determined by the rules governing how to operate with them. These rules, on the other hand, were motivated by the problems that Leibniz wanted to solve. In Wittgenstein's terms, the infinitesimals make sense and have meaning within the sign game developed by Leibniz but are meaningless outside of it. Similarly, a chess figure has no isolated meaning as such, no absolute meaning independent of the whole game and its rules. Meaning always depends on the respective language game or sign game and also reference of the signs to objects will be controlled by the language game. An extreme case in mathematics is the notion of infinite set and infinite cardinal number. It might be difficult for the learner to take a naïve Platonist stance viewing set theory as descriptive of a universe of prefabricated sets (as in Gödel or in Deiser 2010). With Wittgenstein, one can interpret set theory as one possible answer to the question of how one could sensibly talk about infinity. That not every such talk is sensible was shown by the well-known paradoxes. The definitions and the propositions of set theory then are the rules within a language game that develop the grammar of "infinity," and as is known, different such grammars are possible and sensible. Researching the infinite then becomes the more mundane activity of exploring rule systems in regard to their consequences, which is still a wonderful intellectual achievement. The "infinitely large" becomes part of the prose of math. Again we find that it is not some external object (infinite set) that regulates how mathematics is done but mathematics itself that determines how one can view the infinite, which in a way emerges in the respective language game. It should be clear that such views and attitudes bring mathematics back to the purview of human beings, which does not make it any easier to learn but possibly arouses less fear and anxiety about an inaccessible realm far beyond one's reach.

The final notion in Wittgenstein to be mentioned briefly is that of "norm" or "paradigm." In connection with the notions of language game, grammar, and rule use, it permits the dissolution of some of the notorious enigmas ascribed to mathematics: the necessity or unavoidability of mathematics. Mathematics cannot be otherwise and alternatives are not conceivable as is possible for statements, say, about nature. There is no change in mathematics, mathematics is timeless, and its propositions are eternally true and they are exactly true, not only approximately. Furthermore, there is the puzzle of the applicability of mathematics to nature, though the latter is seen as categorically different from mathematics. The way out of many of these enigmas proposed by Wittgenstein is to recognize that mathematical notions and propositions in many cases are used as a norm, as a measuring stick against which something is judged and evaluated. We use established arithmetic rules to judge the correctness of calculations and of counting: Only what conforms to the rules is considered to be acceptable. Those arithmetic propositions and relations are not used as descriptions of eternally true properties of numbers but as templates to carry out and to check the correctness of other calculations, even if the prose tells us otherwise. The mathematical circle, or the mathematical sphere in this sense, is not used as an object but again as a rule to which something to be called a circle or a sphere has to conform. Those uses of math are not descriptive but rather prescriptive or evaluative. Again as with rules, norms or paradigms have no truth value and all the conundrums about mathematical objects for them simply do not make sense. It is the use made of mathematics that makes it timeless, eternal, apodictic, necessary, and, in a trivial sense, true, since that truth results from accepting something as a rule, a norm, or a paradigm. Mathematical propositions are not used as descriptions of facts but are used as rules for description. There is therefore no need to ascribe to them or to mathematical objects any ontological status, since their "reality" resides in their uses within the sign games of mathematics. At least this Wittgenstein would very likely agree with.

4.5 Conclusion

The main purpose of this contribution is to arouse more interest in the views on mathematics and mathematical activity proposed by Peirce and especially Wittgenstein, whose ideas were often overlooked within math education. For some possible consequences of Wittgensteinian ideas for learning mathematics, see Dörfler (2014). A common theme for both of these men is that human intellectual and linguistic activity is fundamentally based on signs of all sorts, and this applies all the more to mathematics. The signs are not just a means or a tool for mathematical activity and creativity, but they are essential and constitutive for mathematics, its notions, and propositions and their meanings. Thus for Peirce, to learn mathematics would be to acquire expertise in diagrammatic reasoning, and for Wittgenstein, it would be to participate in the many various sign games and their techniques. In both cases, which are closely related, it is of great importance to stick meticulously to established rules. This holds for pure mathematics and its proof techniques and for the manifold ways of applying mathematics to other fields. Importantly, mathematics is thereby fundamentally shown to be a deeply social and socially shared cultural activity and product: sign activity can be executed with others and shown to others in a public form. This is very different from imagining mathematics as a kind of abstract and mental activity.

Chapter 5
Networking of Theories in the Tradition of TME

Angelika Bikner-Ahsbahs

5.1 The Networking of Theories Approach

During the last decade, the discussion on theory development has been reconsidered, for example, by Prediger (2010), in the networking of theories approach worked out by a group of European researchers[1] coming from Germany, Italy, Spain, Israel, France, and (at the beginning also from) the UK. The growing complexity of the field and an increase of the diversity of theories motivated this work (see Dreyfus 2009). The aim was to find a scientifically based way of dealing with different theories in research. The current state of the art on the networking of theories approach has been published in a recent book, illustrated by empirical case studies (Bikner-Ahsbahs et al. 2014), and in methodological articles (Kidron and Bikner-Ahsbahs 2012; Bikner-Ahsbahs and Prediger 2010; Prediger et al. 2008; Dreyfus 2009).

The idea of the networking of theories is based on four assumptions (Bikner-Ahsbahs 2009):

1. Regarding the diversity of theories as a form of scientific richness,
2. Acknowledging the specificity of theories,
3. Looking for the connectivity of theories and research results,
4. Developing theory and theory use to inform practice. (p. 7).

[1]Members were and still are: Angelika Bikner-Ahsbahs (speaker), Michèle Artigue, Ferdinando Arzarello, Marianna Bosch, Agnès Corblin-Lenfant, Tommy Dreyfus, Josep Gascón, Ivy Kidron, Stefan Halverscheid, Mariam Haspekian, Susanne Prediger, Kenneth Ruthven, Cristina Sabena, and Ingolf Schäfer.

A. Bikner-Ahsbahs (✉)
Faculty of Mathematics and Information Technology, University of Bremen,
Bibliothekstrasse 1, 28359 Bremen, Germany
e-mail: bikner@math.uni-bremen.de

A. Bikner-Ahsbahs et al., *Theories in and of Mathematics Education*,
ICME-13 Topical Surveys, DOI 10.1007/978-3-319-42589-4_5

The networking of theories allows for explicitly working with different theories in order to benefit from their theoretical strengths with a specific focus on informing practice as well as being inspired by empirical situations of practice. To network theories means to build relations among theories. This approach is not a new idea. There are forerunners: for example, in 1992, Bauersfeld presented an integrated analysis of a teaching and learning situation using various theoretical approaches (1992b). He has also used strategies to compare and contrast radical constructivism and activity theory in order to clarify their (in-) compatibility (1992a). In 1998, Maier and Steinbring published a comparison of two theoretical approaches on processes of understanding based on the same empirical episode. These examples show that German researchers used the strategies of comparing theoretical approaches and integrating theoretical aspects to comprehend practice. Such strategies, called *networking strategies*, have been systematized, resulting in a landscape of four pairs of complementary strategies (Fig. 5.1). This landscape orders these strategies according to their potential for integration between two poles, non-relation of ignoring other theories, and unification of theories globally.

The first two pairs (understanding and making understandable, comparing and contrasting), acknowledge the theories' identities and the diversity of the theories as a resource in the field. They point to the basic necessity of understanding theories. This first pair may take place in a deepened way while comparing and contrasting theories. The second pair, comparing and contrasting, leads to awareness of differences and commonalities, thus learning from other perspectives. The third pair, combining and coordinating, is a step towards bridging theories. Its strategies allow working with different theories to build a conceptual framework or include complementary views into researching a problem. The fourth pair, locally integrating and synthesizing, leads to more comprehensive theoretical frameworks. Local integration may occur when a concept can be interpreted from different theoretical views, thus integrating the concept into other theories. Synthesizing is meant when two or more theories can be imbedded into a more holistic theoretical framework. While there are some cases of local integration, a case of synthesizing has not yet been achieved (see Bikner-Ahsbahs and Prediger 2010).

During the last decade, research methods using networking of theories have been developed in a number of projects. These methods encompass repeated exchanges

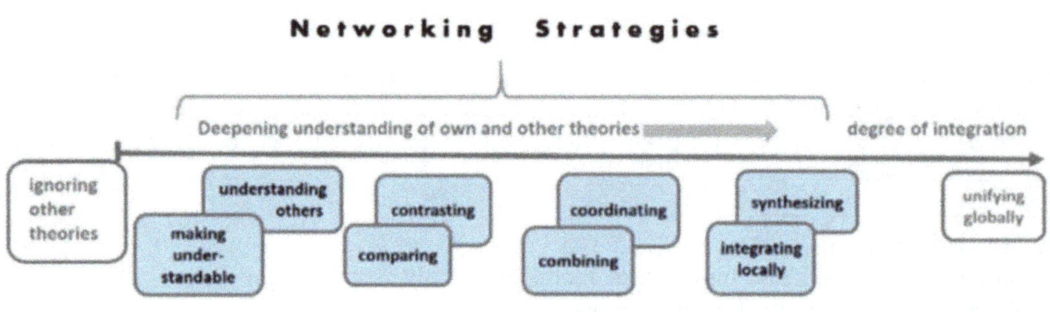

Fig. 5.1 Networking strategies (Prediger et al. 2008, p. 170; Bikner-Ahsbahs and Prediger 2010, p. 492)

of experiments or analyses from various views, using networking strategies to connect theories, establishing a common ground, doing complementary analyses, implementing inclusive methodologies, and producing common methodological and theoretical reflections (Bikner-Ahsbahs 2009; see Bikner-Ahsbahs and Kidron 2015; Artigue and Mariotti 2014).

In contrast to TME, the foundation of a scientific discipline is not directly addressed in the networking of theories approach. Contribution to such a foundation is done by research connecting theories to solve problems and by additional meta-research of final reflections on methodologies and research results. For example, in Kidron et al. (2014), the notions of context were compared and contrasted in the theories "Abstraction in Context" (AiC), "Theory of Didactic Situations" (TDS), and "Anthropological Theory of the Didactic" (ATD). Since all three theories share an epistemological sensitivity, the comparison of the three relevant concepts, *context* in AiC, *milieu* in TDS and *dialectic media-milieu* in ATD, was related to the epistemological nature of the theories and made it possible to sharpen the notions of the three contextual concepts by comparing and contrasting their role in the theories. More generally, this case study showed three ways in which such a broad concept as "context" can be theoretically specified: as a core concept in TDS, as a developmental concept to answer the question of how media and milieu are interrelated in ATD, and as a variable counterpart of the theoretical core of abstraction processes in AiC.

Empirical research within the networking approach has also led to new kinds of theoretical concepts lying at the boundary of theories. These boundary concepts (see Akkermann and Bakker 2011) make sense from different theoretical views, mainly in complementary ways. For example, the "epistemological gap" (Sabena et al. 2014)[2] is a phenomenon that may appear when the epistemic view of the teacher and that of the students differ in that the students do not have access to the same epistemological resources as the teacher. In their example, a student explored the graph of the exponential function and how it develops for big x towards infinity. Using an asymptotic gesture, he described the way the slope of the graph increases towards infinity. His resource consisted of observing the graph on the computer screen; the teacher, however, asked for a proof of contradiction for the statement that the graph of the function could not have an asymptote. The student's epistemological view was based on perception and the teacher's epistemological view was based on logical arguments. They shaped an epistemological gap that the student could not bridge by himself. Two theoretical concepts were used in the analysis, the *semiotic bundle concept* and the concept of *interest-dense situation*. Both concepts integrated into a common methodology allowed for characterising and identifying this epistemological gap (Sabena et al. 2014).

[2]From a theoretical point of view, Burscheid and Struve (1988) have already described such an epistemological (teaching-learning) gap as the gap between the empirical theory of the students and a more logical theoretical background of the teacher.

Networking of theories has been developed as a research practice to solve problems that are so complex that more than one theory should be considered. It tries to handle the complexity of teaching and learning in research and to avoid simplistic interpretations. In contrast to TME, the networking of theories builds on concrete research taking into account different theoretical perspectives, reflecting the methodological processes and the epistemological basis.

Advancing the field is not an explicit issue of the networking of theories, but it can be a result of research conducted through a networking of theories approach. In this way, advancing the field does not happen in big steps but as a very slow process, layer by layer, based on research in the very same research process. As Artigue and Bosch (2014) have outlined, the meta-knowledge gained has not yet reached the level of theoretical or meta-theoretical knowledge so far; it is more a kind of craft knowledge enriched by methodological strategies. These strategies encompass meta-research, which clarifies theories and their assumptions, phenomena and concepts. Thus, the networking of theories further develops theories and the scientific dialogues that take place between researchers from distinct theory cultures (Kidron and Monaghan 2012).

5.2 The Networking of Theories and the Philosophy of the TME Program

The networking of theories was already foreshadowed in a discussion in the Topic Area on TME of ICME 5, just before the first TME conference, summarized by Steiner (1986):

> The ensuing discussion was basically concerned with a comparison of theoretical positions represented by two contributions concentrating on commonalities and differences between "information" and "knowledge."... In general, it was agreed that confronting and comparing different methods for interpretation and analysis of phenomena and problems in mathematics education is a worthwhile task and one to be worked at more intensely in future activities of TME. (p. 296)

Moving to the present day, the networking of theories approach has gone much further. It has developed strategies of *meta-research* building on the research itself as an additional research practice. Such strategies do not observe the field to identify basic problems to be addressed, and they do not offer big lines of theory development, but they do add a deep methodological reflection addressing complexity in the research practice. The new knowledge that has been produced consists of tiny but sustainable steps. In line with Bigalke and Steiner, it respects the *diversity of theories* as richness in the field. Thus, the disciplinary matrix of Kuhn and Masterman and specifically their developmental phases are not considered to be a suitable model for the field. Contextual and hence theoretical diversity, in the sense of networking of theories, provides rich insight mainly into research practices. The networking of theories approach may further develop theories and theory concepts and in this way advance the field. A *unifying paradigm* is explicitly excluded. Since this approach is a methodological and practical one leading to new kinds of concepts at the boundary of theories but also to new kinds of questions addressing complementarity, the advancement of the field may be reached through dialogue (Kidron and Monagham 2012). Research is not restricted to *home-grown theories* but research by networking of theories may develop the field in a *home-grown way*.

What is the potential to advance the field in the sense of TME if we practice networking of theories as a normal research practice? The reader is invited to engage in a networking of theories case and reflect on issues of TME. The example case will be on learning fractions. It will be analysed from the two theoretical perspectives presented before. The two analyses will then be networked to clarify the complementary nature of the two theories.

5.3 An Example of Networking the Two Theoretical Approaches

In a sixth grade class (partly presented in Bikner-Ahsbahs 2005, pp. 234–243, see Bikner-Ahsbahs 2001) the teacher implemented the following task to introduce the concept of fractions for the first time, giving the students three equal bars of chocolate represented as rectangles:

> Four students want to distribute three equal bars of the same chocolate in a fair way among them. How do they manage it? Find at least one distribution.

The students were supposed to work in groups and present their solutions to the class afterwards. Some of the distributions are shown in Figs. 5.2, 5.3, 5.4, 5.5, 5.6 and 5.7 (each student is represented by a different pattern).

In the class, a discussion about sameness and fairness took place. For example: Does everyone get the same in the distribution of Fig. 5.4 even though one gets three small pieces while the others get only one bigger piece each? In a similar way, sameness was discussed for the distributions in Figs. 5.5 and 5.7. The first implicit

Fig. 5.2 Distribution 1

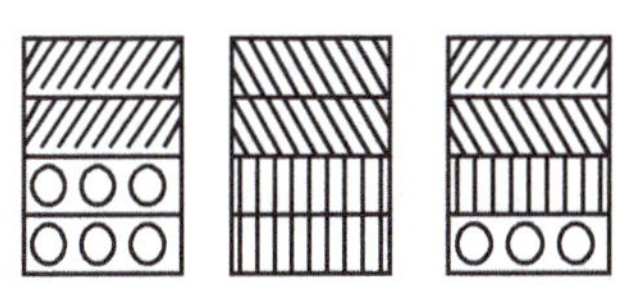

Fig. 5.3 Distribution 2

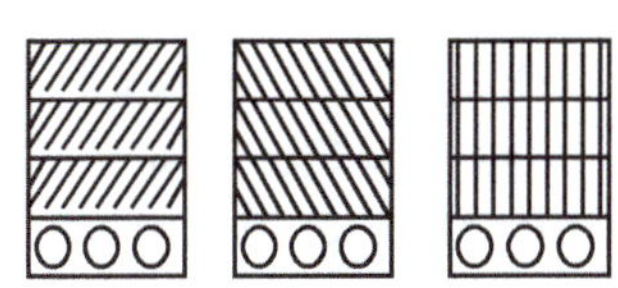

Fig. 5.4 Distribution 3

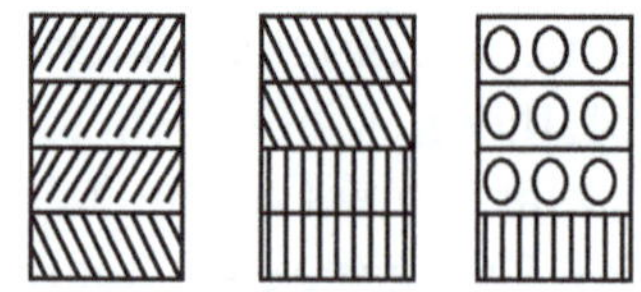

Fig. 5.5 Distribution 4

Fig. 5.6 Distribution 5

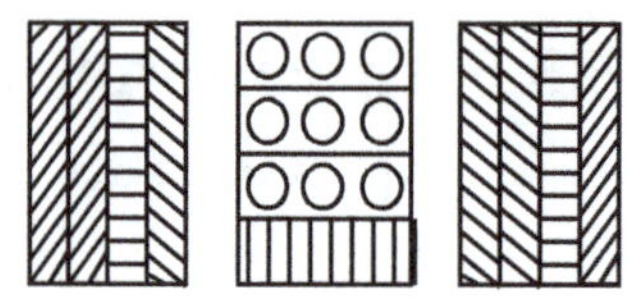

Fig. 5.7 Distribution 6

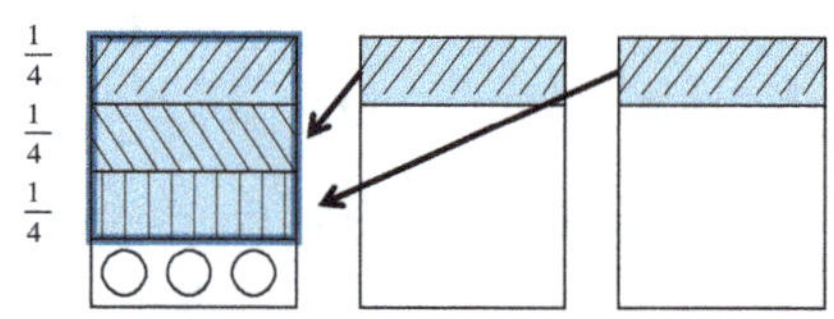

Fig. 5.8 Substituting to check sameness

rule appeared to be: The pieces are the same when they can be substituted by the others. This was shown by the teacher in the diagram in Fig. 5.8 on the blackboard.

However, in Fig. 5.7 this was difficult to achieve. So the rule was changed to: The pieces are the same when they represent the same amount of chocolate. So why did the piece at the bottom of the second rectangle in Fig. 5.7 show the same amount of chocolate as the long parts in the first and the third rectangle? The answer was quickly found: one quarter of the same bars were always of the same amount no matter what shape the quarters have and how they are positioned. But now another question arose: What does everyone get? Three quarters of one bar? In Fig. 5.4, this seemed right for the parts with stripes, but not for the parts with circles. The latter parts rather were described as "three quarters of three bars," while other students said that they were "one quarter of three bars." This again caused a lively discussion about the question: Are three quarters of one bar the same as one quarter of three bars (and three quarters of three bars)? The subsequent discussion showed emotional engagement. Those students who interpreted the preposition *of* as *taken away from* were convinced that three quarters of one bar was not the same as one quarter of three bars because if just one quarter was taken, it was much less than three quarters. The whole as a variable entity was not yet built. One quarter or three quarters were regarded as pieces of an absolute size and not of a relative size

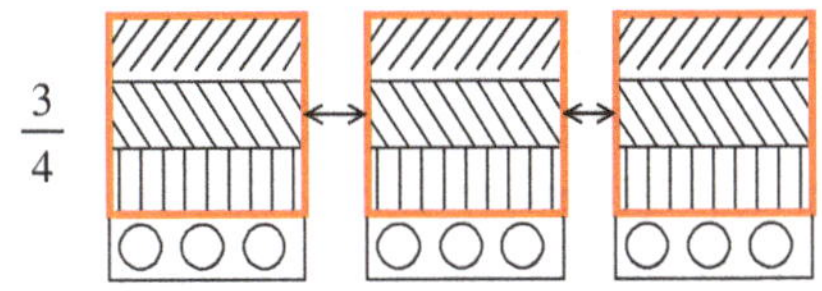

Fig. 5.9 Three quarters of one big bar (three small bars)

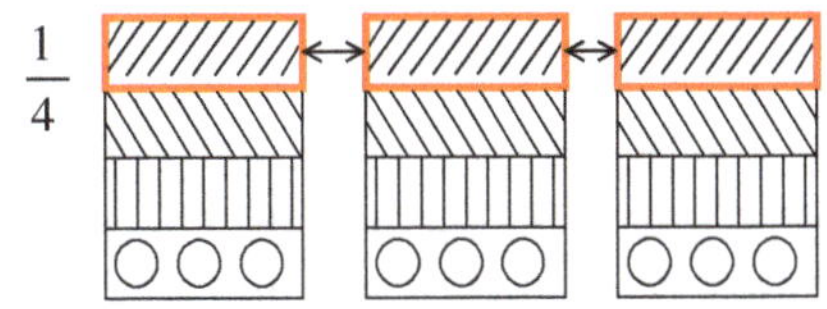

Fig. 5.10 One quarter of one big bar (three small bars)

according to the related entity. Rosa had a nice idea about changing the size of the bar (Bikner-Ahsbahs 2005):

Rosa If we now, if we now join all the three bars together and then we would take from them three *quarters* [emphasized], that would be *too much* [emphasized]. This does not work if one would get three quarters of three bars. (p. 242, translated)

She joined the three bars, getting one big bar (represented by double arrows). Three quarters of this big bar would then be much more than just one quarter of the big bar (Figs. 5.9 and 5.10).

Thus, it became clear that in Fig. 5.4 the part with circles is one quarter of a big bar and that this was the same as three quarters of a small bar, still considering it to be a *part that is taken away*. This was still not acceptable for those students who regarded one quarter as an absolute size. One quarter as a relation between the part and the whole needed further exploration with variable entities, for example varying the size of the whole and investigating what *one quarter of* means.

5.4 The Sign-Game View[3]

The task has initiated an activity by setting the rule to achieve a fair distribution of the chocolate bars represented in the rectangles to be used. The students invented diagrams of distributions showing "the spatial relationships of its parts to one another and the operations and transformations of and with the diagrams" (Dörfler, Sect. 4.3) and inventing the rule that being the same means to be able to substitute the parts (Fig. 5.8). Based on the rule, they used "inventive and constructive manipulation of diagrams to investigate their properties and relationships." (ibid., Sect. 4.3). The students compared their solutions and tried to understand the diagram in a social activity, expressing their interpretations "in natural language and specific terms relating to the diagram," (ibid., Sect. 4.3) such as one quarter or three quarters

[3]We will use quotes to refer to Dörfler's text of Chap. 4 in this book.

in the different figures. "These descriptions and explanations cannot be substituted for the diagram and its various uses, however. In relation to the diagram and its intended relations and operations, this [language] is a meta-language about the diagrams, which also focuses attention and interest on its relevant aspects and activities." (ibid., Sect. 4.3) These aspects and activities consist of the various ways in which three quarters are expressed by diagrams and what they mean compared to each other. However, it also shows that language may result in difficulties; for example, in the question, "Are three quarters of one bar the same as one quarter of three bars and three quarters of three bars?" While the diagrams seemed to be clear, the natural language of the students was not yet conventionalized; hence, the difficulty arose from the differences in the interpretation of what *quarter of/from* means. Exactly this aspect points to another difficulty the students had: regarding one or three quarters as a relationship between the part and the whole. The diagrams presented above do not show this aspect to be relevant. Rosa seemed to be aware of this relationship and invented a way of working with the diagram by changing the size of the whole bar. She began to build the whole as a variable entity by pushing the three bars together to achieve one big bar (Figs. 5.9 and 5.10). It is this action on the diagrams that shows what one quarter of a big bar means, and according to the original rule this is the same as three quarters of one small bar (Figs. 5.9 and 5.10). However, another rule must be added or disclosed by the students: one quarter or three quarters do not have an absolute size but must be used with reference to the whole. Rosa's action shows that "the signs (see Figs. 5.1, 5.2, 5.3, 5.4, 5.5, 5.6, 5.7, 5.8, 5.9 and 5.10) are not just a means or a tool for mathematical activity and creativity, but they are essential and constitutive for mathematics, its notions, and propositions and their meanings" and "sign activity can be executed with others and shown to others in a public form." (ibid., Sect. 4.5). The students are engaged in a "sign game" with the help of the teachers accepting and inventing rules; thus developing mathematical meaning represented in processes of diagramming.

5.5 The Learning Activity View[4]

An activity theory view looks at the students' actions while working on the task. Analyses are normally imbedded in the students' learning biography in school focusing on the current situation and the teacher's intentions and actions. A global view of planning the course of instruction is as relevant as the local view of the ways the teacher supports students in orienting and conducting the task. When choosing the task above, the teacher has to be aware of the cultural historical content which the task allows students to learn by the initiated learning actions. The question to be answered is what knowledge is a prerequisite and, hence, is or should be available.

[4]I thank Regina Bruder and Oliver Schmitt for their assistance in the analysis of the example.

In the given task, the students are supposed to learn the concept of a fraction, which is represented by figurative diagrams. While preparing this task, the teacher should insure that the necessary knowledge is available for carrying out the task, e.g., by implementing calculations for the area of rectangles. The task is supposed to initiate a learning action with the goal of finding out which equal parts of the chocolate bars the four people may get. As a resource, material chocolate bars may be offered and then transformed into iconic representations. This transformation might be introduced by the teacher as a helpful tool allowing mental ways of trying out and manipulating and finally transferring the results back to the real situation again.

Two basic acquisition actions are aimed at *identifying* and *realizing* fractions. First, the students begin to realize the fraction ¾ with the help of diagrams and identify other representations while comparing the students' solutions. The teacher systematizes the students' explorations during the class discourse to assist them in building a pattern or even field orientation for working with similar tasks. This is shown in the discussion about how far three single quarters of the chocolate bar correspond to one piece of ¾ of a bar. The teacher mediates between the knowledge the students have in mind and the knowledge that has been culturally given.

For further instruction, the teacher could use tasks for identifying and realizing fractions in terms of rectangular things or diagrams similar to the ones used before. Solving similar tasks with diagrams of a different shape such as a circle need not be successful based on pattern orientation, but might be a starting point for generalizing the knowledge about representing fractions.

5.6 Comparison of Both Approaches

The sign use approach built on Peirce and Wittgenstein focuses on the slow development of mathematical meaning being situated in manipulating diagrams, its perceivable changes, and diagrammatic reasoning. "This is very different from imagining math as a kind of abstract and mental activity." (Dörfler, in this survey, Sect. 4.5). Although people speak about these diagrams, the mathematical ideas are expressed in the diagrams and not built by mental constructions or images of people. The strength of this approach is its sensitivity towards which diagrams and their development can express mathematical ideas in certain rule-based ways. In the situation above, the part-whole relation of a fraction is diagrammatically unclear since there is only one kind of entity representing the whole. Diagrams and acting with diagrams belong to the kernel of this theoretical approach. Intended applications may consider what kind of acting on diagrams can express which specific mathematical meanings, how language about diagrams is used, and, hence, how sign games can be shaped by people and social groups.

The theory of *learning activity* is also based on acting, but not with a focus on *diagrams to be acted on* but more on the subjects *who are acting on the diagrams as resources to achieve specific cultural knowledge*. Two basic actions are distinguished, identifying and realizing, which can be initiated by tasks. Taking pattern or field orientations into account allows for foreseeing what kinds of tasks the students might solve successfully. Initiating a learning activity does not only focus on the current task situation but requires also taking past learning experience and future goals into account. Tools, e.g., diagrams, do not belong to the kernel of the theory. Its kernel encompasses the concept of activity and how a learning activity can be shaped, initiated by tasks, and created by the learner with the help of the teacher. The teacher's role is crucial. Referring to the example above, one intended application is concerned with the problem of which further tasks the teacher can choose in order to assist the students in building the concept of ¾ to be represented by various shapes. The strength of this approach is its prescriptive nature for initiating learning activities, while diagrams may serve one kind of resource among others.

While both approaches share the sensitivity towards acting, the core concepts (e.g., diagram) of the one theory lie more in the periphery of the other (e.g., as a resources for a learning activity). If we take a networking of theories view and coordinate the analyses by using the two theoretical views, the empirical situation presented may be investigated according to two complementary questions: (1) what and how can acting with diagrams express mathematical ideas and (2) how can a task with certain goals be designed to initiate basic actions, such as identifying and realizing in a specific stage of the course of instruction, that are built on prior knowledge and preparing future goals to achieve cultural knowledge. Thus, both approaches complement each other and may enrich each other to inform practice (see TME program): coming from the learning activity we may zoom into (see Jungwirth 2009 cited by Prediger et al. 2009, p. 1532) diagram use, and coming from diagram use we may zoom out (ibid., p. 1532) to embed the diagram use into the whole course of the learning activity.

Chapter 6
Summary and Looking Ahead

Angelika Bikner-Ahsbahs and Andreas Vohns

A vivid discussion about the theory concepts and the role of theory in mathematics education began at the end of the 1970s. This was embedded into a broader discourse about the nature of the "*Didaktik der Mathematik*" ("Didactics of Mathematics"/mathematics education) and its core subject: the mathematics to be learned and taught. In the 1980s, different theory traditions began to develop in research in the German field, while some meta-theoretical considerations emerged from research within specific paradigms. This German discussion was re-addressed during the TME conferences beginning in 1984, where Steiner presented a program (the TME program) for the foundation of mathematics education as a scientific discipline on an international level. The Networking of Theories approach, established in 2006 to deal with the growing diversity of theories in Europe, can be regarded as a "spiritual TME-successor." It had forerunners in the German field: early examples in the German community stem from Bauersfeld, and Steiner has documented that already the TME conferences had provided space for the dialogue about comparing and contrasting theories in the field. Two theory strands with scientific routes in German-speaking traditions were presented. These theoretical approaches were networked in a case of learning fractions to investigate how they could be related. This case shows the theories' complementary nature, providing a micro-view on a specific moment within a larger view on the learning activity.

What can we learn from this survey for the future of teaching and learning mathematics in Germany and internationally?

A. Bikner-Ahsbahs (✉)
Faculty of Mathematics and Information Technology, University of Bremen,
Bibliothekstrasse 1, 28359 Bremen, Germany
e-mail: bikner@math.uni-bremen.de

A. Vohns
Department of Mathematics Education, Alpen-Adria-Universität Klagenfurt,
Sterneckstraße 15, 9020 Klagenfurt, Austria
e-mail: andreas.vohns@aau.at

A. Bikner-Ahsbahs et al., *Theories in and of Mathematics Education*,
ICME-13 Topical Surveys, DOI 10.1007/978-3-319-42589-4_6

- The discussion on suitable theory concepts and how they may be developed in a home-grown way goes on and should be renewed again and again.
- This discussion is deeply interrelated to the nature and the development of mathematics education as a scientific discipline. As the TME program has stressed, the awareness of what mathematics education is about should be raised and kept alive, reconsidering and deliberating relevant topics/problems and relating them to the practice of teaching and learning mathematics, which is ever changing.
- There seems to be a scientific necessity for meta-theoretical considerations, whether within a theory culture or across theory cultures in mathematics education; top down, such as was proposed by the TME program; or bottom up by research with the networking of theories approach. How this practice will go on will depend on the kinds of problems to be explored in the field.
- The two theories presented are not only analysis tools fitting a suitable aim and theory concept, they also have a past history of which the community of mathematics education should be aware—this holds true for many theories in mathematics education.
- It is worthwhile to reconsider ideas from past research in order to learn more about continuity and change in our scientific discipline and the practice of teaching and learning, in each country as well as internationally.

References

List of References

Akkerman, S., & Bakker, A. (2011). Boundary crossing and boundary objects. *Review of Educational Research*, *81*(2), 132–169.

Artigue, M., & Mariotti, M. A. (2014). Networking theoretical frames: The ReMath enterprise. *Educational Studies in Mathematics*, *85*, 329–355.

Bauersfeld, H. (1992a). Activity theory and radical constructivism—what do they have in common and how do they differ? *Cybernetics and Human Knowing*, *1*(2/3), 15–25.

Bauersfeld, H. (1992b). Integrating theories for mathematics education. *For the Learning of Mathematics*, *12*(2), 19–28.

Bauersfeld, H., Otte, M., & Steiner, H.-G. (Eds.). (1984). *Schriftenreihe des IDM: 30/1984. Zum 10jährigen Bestehen des IDM*. Bielefeld: Universität Bielefeld.

Beck, C., & Jungwirth, H. (1999). Deutungshypothesen in der interpretativen Forschung. *Journal für Mathematik-Didaktik*, *20*(4), 231–259.

Beck, C., & Maier, H. (1993). Das Interview in der mathematikdidaktischen Forschung. *Journal für Mathematik-Didaktik*, *14*(2), 147–180.

Beck, C., & Maier, H. (1994). Mathematikdidaktik als Textwissenschaft. Zum Status von Texten als Grundlage empirischer mathematikdidaktischer Forschung. *Journal für Mathematik-Didaktik*, *15*(1/2), 35–78.

Bigalke, H.-G. (1974). *Sinn und Bedeutung der Mathematikdidaktik. ZDM*, *6*(3), 109–115.

Bigalke, H.-G. (1984). Thesen zur Theoriendiskussion in der Mathematikdidaktik. *Journal für Mathematik-Didaktik*, *5*(3), 133–165.

Bikner-Ahsbahs, A. (2001). Eine Interaktionsanalyse zur Entwicklung von Bruchvorstellungen im Rahmen einer Unterrichtssequenz. *Journal für Mathematik-Didaktik*, *22*(3/4), 179–206.

Bikner-Ahsbahs, A. (2005). *Mathematikinteresse zwischen Subjekt und Situation. Theorie interessendichter Situationen - Baustein für eine mathematikdidaktische Interessentheorie*. Hildesheim, Berlin: Verlag Franzbecker.

Bikner-Ahsbahs, A. (2009). Networking of theories—why and how? Special plenary lecture. In V. Durand-Guerrier, S. Soury-Lavergne & S. Lecluse (Eds.), *Proceedings of CERME 6*. Lyon, France. Retrieved August 23, 2010 from http://www.inrp.fr/publications/edition-electronique/cerme6/plenary-01-bikner.pdf. Accessed 23 August 2010.

Bikner-Ahsbahs, A., & Kidron, I. (2015). A cross-methodology for the networking of theories: The general epistemic need (GEN) as a new concept at the boundary of two theories. In A. Bikner-Ahsbahs, CH. Knipping & N. Presmeg (Eds.), *Approaches to Qualitative Methods in Mathematics Education—Examples of Methodology and Methods* (pp. 233–250). New York: Springer.

A. Bikner-Ahsbahs et al., *Theories in and of Mathematics Education*,
ICME-13 Topical Surveys, DOI 10.1007/978-3-319-42589-4

Bikner-Ahsbahs, A., & Prediger, S. (2006). Diversity of theories in mathematics education—How can we deal with it? *Zentralblatt der Didaktik der Mathematik (ZDM)*, *38*, 52–57.

Bikner-Ahsbahs, A., & Prediger, S. (2010). Networking of theories—An approach for exploiting the diversity of theoretical approaches. With a preface by T. Dreyfus and a commentary by F. Arzarello. In B. Sriraman & L. English (Eds.), *Theories of mathematics education: Seeking new frontiers* (Vol. 1, pp. 479–512). New York: Springer.

Bikner-Ahsbahs, A., Artigue, M., & Haspekian, M. (2014). Topaze effect—A case study on networking of IDS and TDS. In A. Bikner-Ahsbahs & S. Prediger (Eds.) and The Networking Theories Group, *Networking of Theories as a Research Practice in Mathematics Education* (pp. 201–221). New York: Springer.

Bikner-Ahsbahs, A., Prediger, S., & The Networking Theories Group (2014). *Networking of theories as a research practice in mathematics education*. New York: Springer.

Brandt, B., & Krummheuer, G. (2000). Das Prinzip der Komparation im Rahmen der Interpretativen Unterrichtsforschung in der Mathematikdidaktik. *Journal für Mathematik-Didaktik*, *21*(23/4), 193–226.

Brückner, A. (2008). 25 Jahre Potsdamer L-S-A-Modell. In E. Vásárhelyi (Eds.), *Beiträge zum Mathematikunterricht 2008. Vorträge auf der 42. GDM Tagung für Didaktik der Mathematik* (pp. 353–356). Münster: WTM-Verlag.

Bruder, R. (2000). Akzentuierte Aufgaben und heuristische Erfahrungen. In W. Herget & L. Flade (Eds.), *Mathematik lehren und lernen nach TIMSS. Anregungen für die Sekundarstufen* (pp. 69–78). Berlin: Volk und Wissen.

Bruder, R. (2003). Konstruieren – auswählen - begleiten. Über den Umgang mit Aufgaben. In G. Becker, H. Ball & R. Bruder (Eds.), *Friedrich-Jahresheft - Aufgaben. Lernen fördern - Selbstständigkeit entwickeln.* Seelze: Friedrich, 12–15.

Bruder, R. (2005). Ein aufgabenbasiertes anwendungsorientiertes Konzept für einen nachhaltigen Mathematikunterricht - am Beispiel des Themas "Mittelwerte". In G. Kaiser & H.-W. Henn (Eds.), *Mathematikunterricht im Spannungsfeld von Evolution und Evaluation* (pp. 241–250). Hildesheim, Berlin: Franzbecker.

Bruder, R. (2008). Vielfältig mit Aufgaben arbeiten. In R. Bruder, T. Leuders und & A. Büchter (Eds.), *Mathematikunterricht entwickeln* (pp. 18–52). Berlin: Cornelsen Scriptor.

Bruder, R. (2010). Lernaufgaben im Mathematikunterricht. In H. Kiper, W. Meints, S. Peters, S. Schlump, & S. Schmit (Eds.), *Lernaufgaben und Lernmaterialien im kompetenzorientierten Unterricht* (pp. 114–124). Stuttgart: W. Kohlhammer Verlag.

Bruder, R., & Brückner, A. (1989). Zur Beschreibung von Schülertätigkeiten im Mathematikunterricht - ein allgemeiner Ansatz. *Pädagogische Forschung*, *30*(6), 72–82.

Bruder, R., & Collet, C. (2011). *Problemlösen lernen im Mathematikunterricht*. Berlin: Cornelsen Scriptor.

Burscheid, H. J. (1983). Formen der wissenschaftlichen Organisation in der Mathematikdidaktik. *Journal für Mathematikdidaktik*, *3*, 219–240.

Burscheid, H. J., & Struve, H. (1988). The epistemological teaching-learning gap. In Alfred Vermandel (Ed.), *Theory of mathematics education. Proceedings of the Third International Theories of Mathematics Education Conference* (pp. 34–45). Antwerpen: Organizing Committee of the Third International TME Conference.

Collet, C., & Bruder, R. (2008). Longterm-study of an intervention in the learning of problem-solving in connection with self-regulation. In O. Figueras, J. L. Cortina, S. Alatorre, T. Rojano, & A. Sepúlveda (Eds.), *Proceedings of the Joint Meeting of PME 32 and PME-NA XXX* (Vol. 2, pp. 353–360). Morelia: Cinvestav-UMSNH.

Davydov, V. V. (1990). *Types of generalization in instruction: Logical and psychological problems in the structuring of school curricula. Soviet studies in mathematics education* (Vol. 2). Reston, VA: National Council of Teachers of Mathematics.

Deiser, O. (2010). *Einführung in die Mengenlehre*. Berlin-Heidelberg: Springer.

Dörfler, W. (2004). Diagrams as means and objects of mathematical reasoning. In H.-G. Weigand (Ed.), *Developments in mathematics education in German-speaking countries. Selected Papers from the Annual Conference on Didactics of Mathematics 2001* (pp. 39–49). Hildesheim: Verlag Franzbecker.

Dörfler, W. (2006). Diagramme und Mathematikunterricht. *Journal für Mathematik-Didaktik, 27* (3), 200–219.

Dörfler, W. (2008). Mathematical reasoning: Mental activity or practice with diagrams. In M. Niss (Ed.), *ICME 10 Proceedings, Regular Lectures, CD-Rom.* Roskilde: IMFUFA, Roskilde University.

Dörfler, W. (2013a). Bedeutung und das Operieren mit Zeichen. In M. Meyer, E. Müller-Hill, & I. Witzke (Eds.), *Wissenschaftlichkeit und Theorieentwicklung in der Mathematikdidaktik.* (pp. 165–182). Hildesheim: Franzbecker.

Dörfler, W. (2013b). Impressionen aus (fast) vier Jahrzehnten Mathematikdidaktik. *Mitteilungen der Gesellschaft für Didaktik der Mathematik,* (95), 8–14.

Dörfler, W. (2014). Didaktische Konsequenzen aus Wittgensteins Philosophie der Mathematik. In H. Hahn (Ed.), *Anregungen für den Mathematikunterricht unter der Perspektive von Tradition, Moderne und Lehrerprofessionalität* (pp. 68–80). Hildesheim: Franzbecker.

Dreyfus, T. (2009). Ways of working with different theoretical approaches in mathematics education research: An introduction. In V. Durand-Guerrier, S. Soury-Lavergne & S. Lecluse (Eds.), *Proceedings of CERME 6.* Lyon, France. http://www.inrp.fr/publications/edition-electronique/cerme6/plenary-01-bikner.pdf. Accessed 18 March 2016.

Duval, R. (1995). *Sémiosis et pensée humaine.* Bern: Peter Lang.

Engeström, Y. (2005). *Developmental work research: Expanding activity theory in practice.* Berlin: Lehmanns Media.

Engeström, Y. (2008). *Entwickelnde Arbeitsforschung: Die Tätigkeitstheorie in der Praxis.* Berlin: Lehmanns Media.

Feldt, N. (2013). Konkretisierung und Operationalisierung von Grundwissen und Grundkönnen durch ein theoriegeleitetes Vorgehen. In G. Greefrath, F. Käpnick, & M. Stein (Eds.), *Beiträge zum Mathematikunterricht 2013* (pp. 308–311). Münster: WTM Verlag.

Fischer, R. (1983). Wie groß ist die Gefahr, daß die Mathematikdidaktik bald so ist wie die Physik? – Bemerkungen zu einem Aufsatz von Hans Joachim Burscheid. *Journal für Mathematikdidaktik, 3,* 241–253.

Fischer, R. (2001). Höhere Allgemeinbildung. In R. Aulke, A. Fischer-Buck & K. Garnitschnig (Eds.), *Situation - Ursprung der Bildung* (pp. 151–161). Norderstedt: Fischer. F. Fischer (Ed.), *Situation - Ursprung der Bildung.* Bd. 6. (pp. 151–161). Leipzig: Wissenschaftsverlag.

Freudenthal, H. (1974). Sinn und Bedeutung der Didaktik der Mathematik. *ZDM, 6*(3), 122–124.

Galperin, P. J. (1967). Die Entwicklung der Untersuchungen über die Bildung geistiger Operationen. In H. Hiebsch (Ed.), *Ergebnisse der sowjetischen Psychologie* (pp. 367–405). Berlin: Akademie-Verlag.

Giest, H., & Lompscher, J. (2006). *Lerntätigkeit - Lernen aus kultur-historischer Perspektive.* Lehmanns Media: Ein Beitrag zur Entwicklung einer neuen Lernkultur im Unterricht. Berlin.

Griesel, H. (1974). Überlegungen zur Didaktik der Mathematik als Wissenschaft. *ZDM, 6*(3), 115–119.

Griesel, H. (2001). Scientific orientation of mathematical instruction—history and chance of a guiding principle in East and West Germany. In H.-G. Weigand (Ed.), *Developments in mathematics education in Germany. Selected papers from the Annual Conference on Didactics of Mathematics Leipzig, 1997* (pp. 75–83). Hildesheim: Franzbecker.

Hersh, R. (1986). Some proposals for reviving the philosophy of mathematics. In T. Tymoczko (Ed.), *New directions in the philosophy of mathematics* (pp. 9–28). Boston: Birkhäuser.

Hoffmann, M. H. G. (Ed.). (2003). *Mathematik verstehen. Semiotische Perspektiven.* Hildesheim: Franzbecker.

Hoffmann, M. H. G. (2005). *Erkenntnisentwicklung.* Klostermann: Ein semiotischer-pragmatischer Ansatz. Frankfurt a. M.

Hoffmann, M. H. G., Lenhard, J., & Seeger, F. (Eds.). (2005). *Activity and sign—grounding mathematics education*. New York: Springer.

Jahnke, H. N. (1978). *Zum Verhältnis von Wissensentwicklung und Begründung in der Mathematik - Beweisen als didaktisches Problem*. Bielefeld: Universität Bielefeld.

Jungwirth, H. (1994). Die Forschung zu Frauen und Mathematik: Versuch einer Paradigmenklärung. *Journal für Mathematik-Didaktik*, *15*(3/4), 253–276.

Kadunz, G. (Ed.). (2010). *Sprache und Zeichen. Zur Verwendung von Linguistik und Semiotik in der Mathematikdidaktik*. Hildesheim: Franzbecker.

Kadunz, G. (Ed.). (2015). *Semiotische Perspektiven auf das Lernen von Mathematik*. Berlin-Heidelberg: Springer Spektrum.

Kaiser (2003, Ed.). Qualitative empirical methods in mathematics education—Discussions and reflections, *Zentralblatt für Didaktik der Mathemaik 35 (5 and 6)*.

Kaiser, G. (2000). Internationale Vergleichsuntersuchungen - eine Auseinandersetzung mit ihren Möglichkeiten und Grenzen. *Journal für Mathematik-Didaktik*, *21*(3/4), 171–192.

Kidron, I., & Bikner-Ahsbahs, A. (2015). Advancing research by means of the networking of theories. In A. Bikner-Ahsbahs, Ch. Knipping, & N. Presmeg (Eds.), *Approaches to qualitative methods in mathematics education—Examples of methodology and methods* (pp. 221–232). New York: Springer.

Kidron, I., & Monaghan, J. (2012). Complexity of dialogue between theories: Difficulties and benefits. In *Pre-proceedings of the 12th International Congress on Mathematical Education*. (pp. 7078–7084). Paper presented in the Topic Study Group 37. COEX, Seoul (Korea): ICME.

Kidron, I., Artigue, M., Bosch, M., Dreyfus, T, & Haspekian, M. (2014). Context, milieu and media-milieus dialectic: A case study on networking of AiC, TDS, and ATD. In A. Bikner-Ahsbahs, S. Prediger and The Networking Theories Group, *Networking of theories as a research practice in mathematics education* (pp. 153–177). New York: Springer.

Kienzler, W. (1997). *Wittgensteins Wende zu seiner Spätphilosophie 1930-1932*. Frankfurt am Main: Suhrkamp.

KMK – Kultusministerkonferenz der Länder (1964/71). *Abkommen zwischen den Ländern der Bundesrepublik zur Vereinheitlichung auf dem Gebiete des Schulwesens. Beschluss der KMK vom 28.10.1964 in der Fassung vom 14.10.1971*. http://www.kmk.org/fileadmin/Dateien/veroeffentlichungen_beschluesse/1964/1964_10_28-Hamburger_Abkommen.pdf. Accessed 10 March 2016.

Knoche, N., & Lind, D. (2000). Eine Analyse der Aussagen und Interpretationen von TIMSS unter Betonung methodologischer Aspekte. *Journal für Mathematik-Didaktik*, *21*(1), 3–27.

Kuhn, T. S. (1970). *The structure of scientific revolutions* (2nd ed.). Chicago: University of Chicago Press.

Ladel, S., & Kortenkamp, U. (2013). An activity-theoretic approach to multi-touch tools in early maths learning. *The International Journal for Technology in Mathematics Education*, *20*, 3–8.

Lompscher, J. (1972). Wesen und Struktur allgemeiner geistiger Fähigkeiten. In J. Lompscher, E.-L. Hischer, L. Irrlitz, W. Jantos, & R. Stahl (Eds.), *Theoretische und experimentelle Untersuchungen zur Entwicklung geistiger Fähigkeiten* (pp. 17–73). Berlin: Volk und Wissen.

Lompscher, J. (1976). *Verlaufsqualitäten der geistigen Tätigkeit*. Berlin: Volk und Wissen.

Lompscher, J. (1985). Die Lerntätigkeit als dominierende Tätigkeit des jüngeren Schulkindes. In J. Lompscher & L. Irrlitz (Eds.), *Persönlichkeitsentwicklung in der Lerntätigkeit* (pp. 23–52). Ein Lehrbuch für die pädagogische Psychologie an Instituten für Lehrerbildung. Berlin: Volk und Wissen.

Maier, H. (1998). „Erklären“: Ziel mathematikdidaktischer Forschung? *Journal für Mathematik-Didaktik*, *18*(2/3), 239–241.

Maier, H., & Beck, C. (2001). Zur Theoriebildung in der Interpretativen mathematikdidaktischen Forschung. *Journal für Mathematik-Didaktik*, *22*(1), 29–50.

Maier, H., & Steinbring, H. (1998). Begriffsbildung im alltäglichen Mathematikunterricht - Darstellung und Vergleich zweier Theorieansätze zur Analyse von Verstehensprozessen. *Journal für Mathematik-Didaktik*, *19*(4), 292–330.

Mann, I. (1990). *Lernen können ja alle Leute. Lesen-, Rechnen-, Schreibenlernen mit der Tätigkeitstheorie.* Weinheim und Basel: Beltz.

Mason, J., & Johnston-Wilder, S. (Eds.). (2004). *Fundamental constructs in mathematics education*. London and New York: Routledge Falmer.

Masterman, M. (1970). The nature of a paradigm. In I. Lakatos & A. Musgrave (Eds.), *Criticism and the growth of knowledge. Proceedings of the International Colloquium in the Philosophy of Science, London, 1965* (pp. 59–90). London: Cambridge.

Mühlhölzer, F. (2010). *Braucht die Mathematik eine Grundlegung? Ein Kommentar des Teil III von Wittgensteins Bemerkungen über die Grundlagen der Mathematik*. Frankfurt: Vittorio Klostermann.

Müller, G. N., & Wittmann, E. (1984). *Der Mathematikunterricht in der Primarstufe: Ziele - Inhalte Prinzipien - Beispiele* (3rd ed.). Wiesbaden: Vieweg + Teubner.

Nitsch, R. (2015). *Diagnose von Lernschwierigkeiten im Bereich funktionaler Zusammenhänge*. Wiesbaden: Springer Spektrum.

Nitsch, R., Fredebohm, A., Bruder, R., Kelava, T., Naccarella, D., Leuders, T., et al. (2015). Students' competencies in working with functions in secondary mathematics education—Empirical examination of a competence structure model. *International Journal of Science and Mathematics Education, 13*(3), 657–682.

Otte, M. (1974). *Didaktik der Mathematik als Wissenschaft. ZDM, 6*(3), 125–128.

Otte, M. (1997). Mathematik und Verallgemeinerung - Peirce' semiotisch-pragmatische Sicht. *Philosophia naturalis, 34*(2), 175–222.

Otte, M. (2011). Evolution, learning, and semiotics from a Peircean point of view. *Educational Studies in Mathematics, 77*(2), 313–322.

Peirce, CH. S. (1931–1958). *Collected papers* (Vol. I–VIII). Cambridge: Harvard University Press.

Pippig, G. (1985). *Aneignung von Wissen und Können - psychologisch gesehen*. Berlin: Volk und Wissen.

Prediger, S., Bikner-Ahsbahs, A., & Arzarello, F. (2008). Networking strategies and methods for connecting theoretical approaches: first steps towards a conceptual framework. *ZDM-International Journal on Mathematics Education, 40*(2), 165–178.

Prediger, S., Bosch, M., Kidron, I., Monaghan, J., & Sensevy, G. (2009). Introduction to the working group 9. Different theoretical perspectives and approaches in mathematics education research – strategies and difficulties when connecting theories. In *Proceedings of CERME 6*. Lyon, France. http://ife.ens-lyon.fr/publications/edition-electronique/cerme6/wg9-00-introduction.pdf. Accessed 18 February 2016.

Radford, L., Schubring, G., & Seeger, F. (Eds.). (2008). *Semiotics in mathematics education: Epistemology, history, classroom, and culture*. Rotterdam: Sense Publishers.

Rotman, B. (2000). *Mathematics as sign: Writing, imagining, counting*. Stanford: Stanford University Press.

Rückriem, G., Giest, H. (2006). Nachruf auf Joachim Lompscher. In H. Giest (Ed.), Erinnerung für die Zukunft. *Pädagogische Psychologie in der DDR* (pp. 159–164). Berlin: Lehmanns Media.

Sabena, C., Arzarello, A., Bikner-Ahsbahs, A., & Schäfer, I. (2014). The epistemological gap—A case study on networking of APC and IDS. In A. Bikner-Ahsbahs & S. Prediger (Eds.) and the Networking Theories Group, *Networking of theories as a research practice in mathematics education* (pp. 165–183). New York: Springer.

Schmitt, O. (2013). Tätigkeitstheoretischer Zugang zu Grundwissen und Grundkönnen. In G. Greefrath, F. Käpnick, & M. Stein (Eds.), *Beiträge zum Mathematikunterricht 2013* (pp. 894–897). Münster: WTM-Verlag.

Shapiro, S. (2000). *Thinking about mathematics. The philosophy of mathematics*. Oxford: Oxford University Press.

Sill, H.-D., & Sikora, Ch. (2007). *Leistungserhebungen im Mathematikunterricht: Theoretische und empirische Studien*. Hildesheim: Franzbecker.

Siller, H.-S., Bruder, R., Hascher, T., Linnemann, T., Steinfeld, J. & Sattlberger, E. (2015). Competency level modelling for school leaving examination. In Konrad Krainer & Nad'a Vondrovà (Eds.), *Proceedings of the CERME 9*. Prague: Charles University. http://files.cerme9.org/200000288-e48a3e582b/TWG%2017%2C%20collected%20papers.pdf. Accessed 13 April 2016.

Skovsmose, O. (1989). Models and reflective knowledge. In *ZDM*, *21*(3)1, 3–8.

Steiner, H.-G. (1983). Zur Diskussion um den Wissenschaftscharakter der Mathematikdidaktik. *Journal für Mathematik-Didaktik*, *3*, 245–251.

Steiner, H.-G. (1985). Theory of mathematics education (TME): An introduction. *For the Learning of Mathematics*, *5*(2), 11–17.

Steiner, H.-G. (1986). Topic areas: Theory of mathematics education (TME). In Marjorie Carss (Ed.), *Proceedings of the Fifth International Congress on Mathematical Education* (pp. 293–299). Boston, Basel, Stuttgart: Birkhäuser.

Steiner, H.-G. (1987a). A systems approach to mathematics education. *Journal for Research in Mathematics Education*, *18*(1), 46–52.

Steiner, H.-G. (1987b). Philosophical and epistemological aspects of mathematics and their intersection with theory and practice in mathematics education. *For the Learning of Mathematics*, *7*(1), 7–13.

Steiner, H.-G. (1987c). Implication for scholarship of a theory of mathematics ecucation. *Zentralblatt der Didaktik der Mathematik, Informationen*, *87*(4), 162–167.

Steiner, H.G., Balacheff, N., Mason, J., Steinbring, H., Steffe, L.P., Cooney, T.J., & Christinasen, B. (1984). Theory of mathematics education (TME). *ICME 5—Topic Area and Mini Conference. Occasional Paper 54, Arbeiten aus dem Institut für Didaktik der Mathematik der Universität Bielefeld.* Bielefeld: IDM. http://www.uni-bielefeld.de/idm/serv/dokubib/occ54.pdf. Accessed 08 April 2016.

Stjernfelt, F. (2000). Diagrams as centerpiece of a peircean epistemology. *Transactions of the C.S. Peirce Society*, *36*(3), 357–392.

Toepell, M. (2004). Zur Gründung und Entwicklung der Gesellschaft für Didaktik der Mathematik (GDM). *Mitteilungen der GDM*, *78*, 147–152.

Vom Hofe, R. (1995). *Grundvorstellungen mathematischer Inhalte*. Heidelberg: Spektrum.

Walsch, W. (2003). Methodik des Mathematikunterrichts als Lehr- und Wissenschaftsdisziplin. *ZDM*, *35*(4), 153–156.

Wellenreuther, M. (1997). Hypothesenbildung, Theorieentwicklung und Erkenntnisfortschritt in der Mathematikdidaktik: Ein Plädoyer für Methodenvielfalt. *Journal für Mathematik-Didaktik*, *18*(2/3), 186–216.

Wittgenstein, L. (1999). *Bemerkungen über die Grundlagen der Mathematik*. Werkausgabe (Vol. 6). Frankfurt: Suhrkamp.

Wittmann, E. C. (1974). Didaktik der Mathematik als Ingenieurwissenschaft. *ZDM*, *6*(3), 119–121.

Wittmann, E. C. (1995). Mathematics education as a 'design science'. *Educational Studies in Mathematics*, 355–374.

Woolfolk, A. (2008). *Pädagogische psychologie*. München: Pearson Studium.

List of References for Further Reading

Batanero, M. C., Godino, J. D., Steiner, H. G., & Wenzelburger, E. (1992). An international TME survey: Preparation of researchers in mathematics education. *Occasional Paper 135, Arbeiten aus dem Institut für Didaktik der Mathematik der Universität Bielefeld.* Bielefeld: IDM.

Bauersfeld, H. (1992). Integrating theories for mathematics education. *For the Learning of Mathematics*, *12*(2), 19–28.

Bikner-Ahsbahs, A., & Prediger, S. (2010). Networking of theories—An approach for exploiting the diversity of theoretical approaches; with a preface by T. Dreyfus and a commentary by F. Arzarello. In B. Sriraman & L. English (Eds.), *Theories of mathematics education: Seeking new frontiers* (Vol. 1, pp. 479–512). New York: Springer.

Bikner-Ahsbahs, A., Prediger, S. & The Networking Theories Group (2014). *Networking of theories as a research practice in mathematics education*. New York: Springer.

Giest, H., & Lompscher, J. (2003). Formation of learning activity and theoretical thinking in science teaching. In A. Kozulin, B. Gindis, V. S. Ageyev & Suzanne M. Miller (Eds.), *Vygotsky's educational theory in cultural context* (pp. 267–288). Cambridge: Cambridge University Press.

Lompscher, J. (1999a). Activity formation as an alternative strategy of instruction. In Y. Engeström, R. Miettinen & R.-L. Punamäki (Eds.), *Perspectives on activity theory* (pp. 264–281). Cambridge: Cambridge University Press.

Lompscher, J. (1999b). Learning activity and its formation: Ascending from the abstract to the concrete. In M. Hedegaard & J. Lompscher (Eds.), *Learning activity theory* (pp. 139–166). Aarhus: Aarhus University Press.

Lompscher, J. (2002). The category of activity—A principal constituent of cultural-historical psychology. In D. Robbins & A. Stetsenko (Eds.), *Vygotsky's psychology: Voices from the past and present* (pp. 79–99). New York: Nova Science Press.

Steiner, H. G., Balacheff, N., Mason, J., Steinbring, H., Steffe, L. P., Cooney, T. J. & Christiansen, B. (1984). Theory of mathematics education (TME). *ICME 5 – Topic Area and Mini Conference. Occasional Paper 54, Arbeiten aus dem Institut für Didaktik der Mathematik der Universität Bielefeld.* Bielefeld: IDM. http://www.uni-bielefeld.de/idm/serv/dokubib/occ54.pdf. Accessed 08 April 2016.

Steiner, H.-G. & Vermandel, A. (1988, Eds.). *Foundations and methodology of the discipline mathematics education. Didactics of mathematics. Proceedings of the Second TME-Conference*. Antwerp: University of Antwerp.

Steiner, H.-G. (1984). Topic areas: Theory of mathematics education. In Marjorie Carss (Ed.), *Proceedings of the Fifth International Congress on Mathematics Education* (pp. 293–298). Boston, Basel, Stuttgart: Birkhäuser.

Vermandel, A. (1988). *Theory of mathematics education. Proceedings of the Third International Conference.* Antwerp: University of Antwerp.

www.ingramcontent.com/pod-product-compliance
Lightning Source LLC
LaVergne TN
LVHW082240200926
848484LV00008B/2380

* 9 7 8 1 0 1 3 2 6 7 5 1 2 *